Caterina Longo • Giuseppe Argenziano
Aimilios Lallas • Elvira Moscarella
Simonetta Piana

Atlas of Diagnostically Challenging Melanocytic Neoplasms

Springer

Caterina Longo
Dermatology and Skin Cancer Unit
Arcispedale Santa Maria Nuova-IRCCS
Department of Dermatology
University of Modena and Reggio Emilia
Reggio Emilia
Italy

Aimilios Lallas
First Department of Dermatology
Aristotle University
Thessaloniki
Greece

Simonetta Piana
Pathology Unit
Arcispedale Santa Maria Nuova-IRCCS
Reggio Emilia
Italy

Giuseppe Argenziano
Dermatology Unit
University of Campania
Naples
Italy

Elvira Moscarella
Dermatology and Skin Cancer Unit
Arcispedale Santa Maria Nuova-IRCCS
Reggio Emilia
Italy

ISBN 978-3-319-48651-2 ISBN 978-3-319-48653-6 (eBook)
https://doi.org/10.1007/978-3-319-48653-6

Library of Congress Control Number: 2017955686

Printed on acid-free paper

This Springer imprint is published by Springer Nature
The registered company is Springer International Publishing AG
The registered company address is: Gewerbestrasse 11, 6330 Cham, Switzerland

Preface

I must definitely applaud the original idea of Caterina Longo of producing this atlas, which is a superb collection of cases outlining the diagnostic challenges we are facing every day with melanocytic tumors. For clinicians managing patients with melanoma, it is clear-cut that differentiating melanoma from benign melanocytic lesions can be very difficult at times. In the last 20 years, the practice changed dramatically because of the introduction of new tools for the preoperative diagnosis. Especially the introduction of dermoscopy and confocal microscopy allowed clinicians to improve their ability to recognize the many faces of melanocytic tumors with a quasi-histopathologic accuracy. However, there is a group of lesions that still are difficult to diagnose because of their equivocal clinical and/or histopathologic features. In these cases, only a careful clinic-pathologic correlation is the method to rich more closely the final diagnosis and, thus, the correct patient management. What I learned in the last 20 years is that nobody has 100% diagnostic accuracy, neither the clinician nor the pathologist. In difficult cases only an open-minded discussion among clinicians and pathologists is able to make the difference. Only by integrating all the possible information, the history, the clinical data, and the histopathologic features might we reach a more reasonable management of our patients. The aim of this book is indeed to illustrate this method, the clinic-pathologic correlation of difficult cases!

Naples, Italy Giuseppe Argenziano, M.D., Ph.D.

Contents

1 Flat Solitary Pigmented Lesions in the Elderly

The sentence "we are born and we will die without nevi" summarizes one of the key components of the diagnosis of atypical lesions in the elderly. Epidemiologic data demonstrate that the nevus count and prevailing nevus patterns are strongly influenced by age. Notably, nevus count increases from childhood to midlife and decreases thereafter. In light of these findings, if evolving nevi in adolescence are an expected finding and therefore do not require further interventions, a melanocytic skin lesion showing signs of growth in the elderly should raise the index for malignancy. Furthermore, any flat acquired melanocytic lesion in this age should be considered with caution since the majority of lesions in the elderly are persistent intradermal nevi (congenital type). Firstly termed as atypical lentiginous junctional melanocytic proliferations, indeed they are regarded nowadays as melanomas.

Clinically, these atypical lentiginous junctional melanocytic proliferations of the elderly are commonly located on the upper back, shoulders, or extremities. They are solitary, often large (>8 mm), ill-defined macules with different shades of black, brown, and gray. Dermoscopically, these lesions are typified by a more or less atypical pigmented network, diffuse structureless brown pigmentation, and areas of regression. Studies employing digital dermoscopic follow-up suggest that these melanomas belong to a group of slow-growing tumors, which may grow in situ for several years. Their histologic diagnosis can be very difficult as they are cytologically bland and show scant epidermotropism and very little atypia. In the early phases, only a numerical increase of the melanocytes, scattered along the basal epidermis, can be noted. The alternation of single cells and irregular nests, with skip areas, usually within a sun-damaged skin, is an important histological clue for a diagnosis of early in situ melanoma.

What is a common finding in this age group is the presence of several benign non-melanocytic skin lesions such as seborrheic keratosis, angiomas, or solar lentigos. Thus, the clues to identify flat melanomas are the following: solitary flat pigmented lesions, large size, with network and regression on dermoscopy. Conversely, the presence of rough surface, comedo-like openings, fingerprinting, and red lacunae should be regarded as benign clues. However, the recognition of incipient melanomas should always be based on clinical data, patient's phenotype, and history.

This section depicts the clinical and dermoscopic features of common and problematic melanocytic tumors in the elderly while trying to provide clues and rules for the correct diagnosis.

C. Longo et al., *Atlas of Diagnostically Challenging Melanocytic Neoplasms*,
https://doi.org/10.1007/978-3-319-48653-6_1

Case 1

Patient: Female, 62 years old

Anatomic site of the lesion: Right leg

History of the lesion: Acquired lesion, present since a few years

Management: Although the dermoscopic aspect revealed the presence of few criteria for melanoma diagnosis, the lesion was excised to rule out melanoma diagnosis since it was a larger and darker lesion in the context of the patient's other nevi.

First histopathologic diagnosis: Junctional nevus

Second histopathologic diagnosis after clinicopathologic consultation: Junctional nevus versus early melanoma in situ. The latter hypothesis seemed the most likely.

Key message: Slow-growing melanomas can show very bland cytological features. The correlation among clinical data (in this case, patient's age), dermoscopic characteristics, and histological features is mandatory.

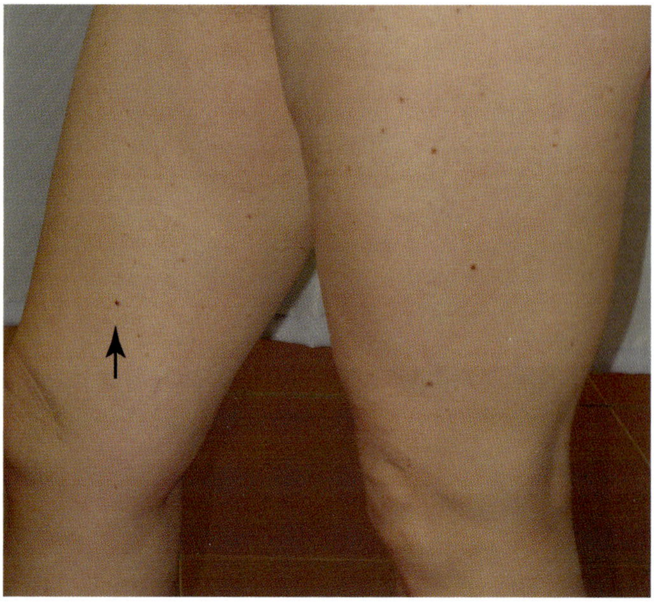

Fig. 1.1 Flat pigmented lesion in the context of a patient with multiple small-sized moles

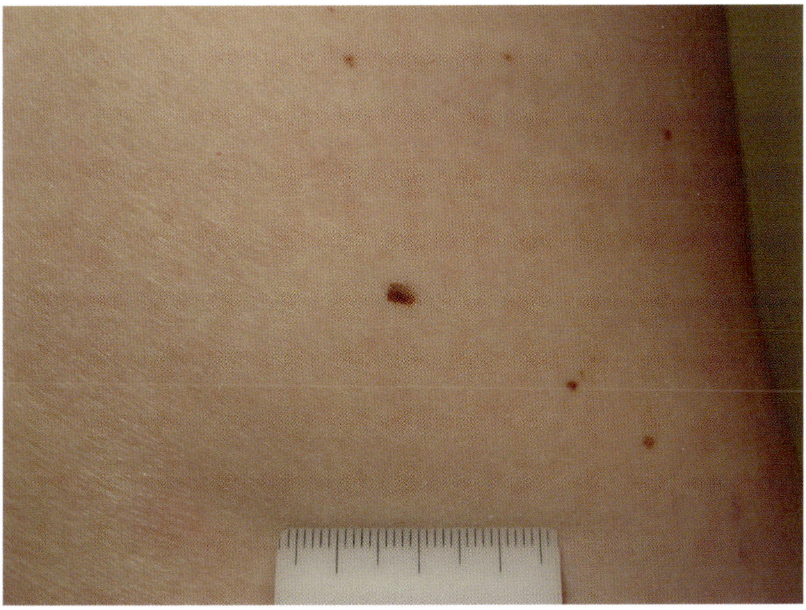

Fig. 1.2 Flat pigmented macule and darker compared to the surrounding lesions with irregular borders

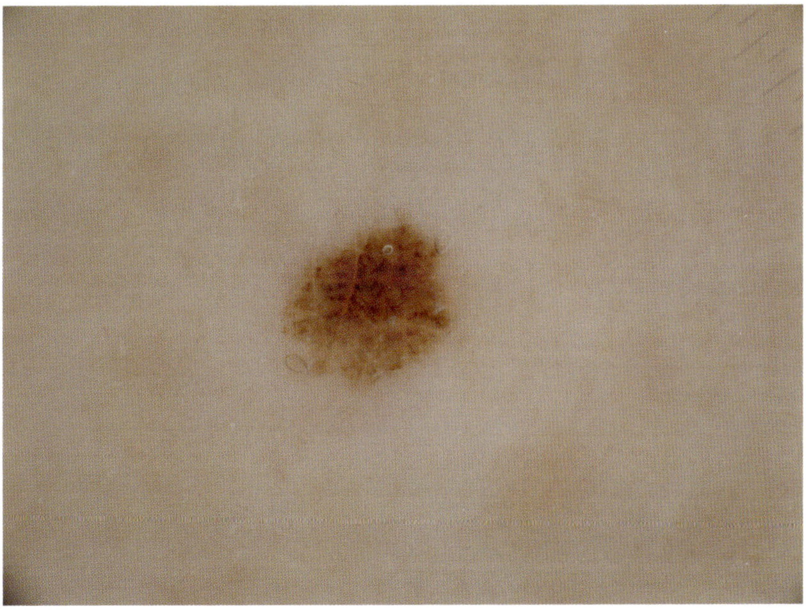

Fig. 1.3 Dermoscopically the lesion was typified by the presence of atypical pigmented network, with the presence of wider skin markings

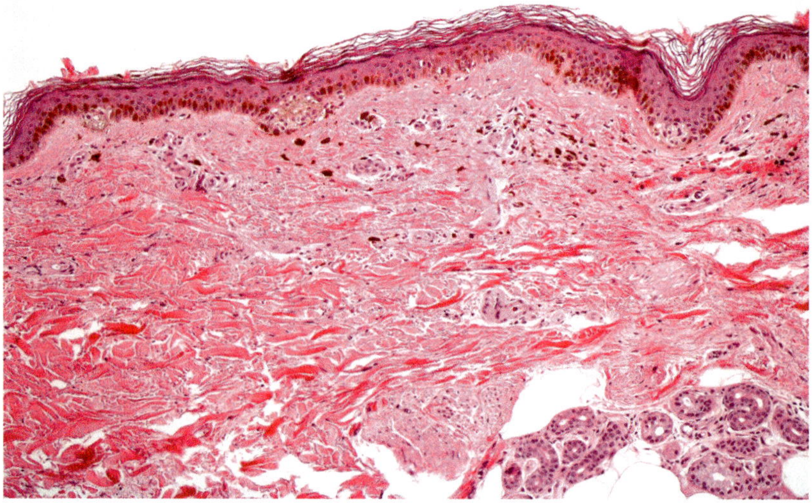

Fig. 1.4 On histology, the lesion was a very bland junctional melanocytic proliferation, mainly arranged in single cells or forming little nests. Neither marked cellular atypia nor epidermotropism was present

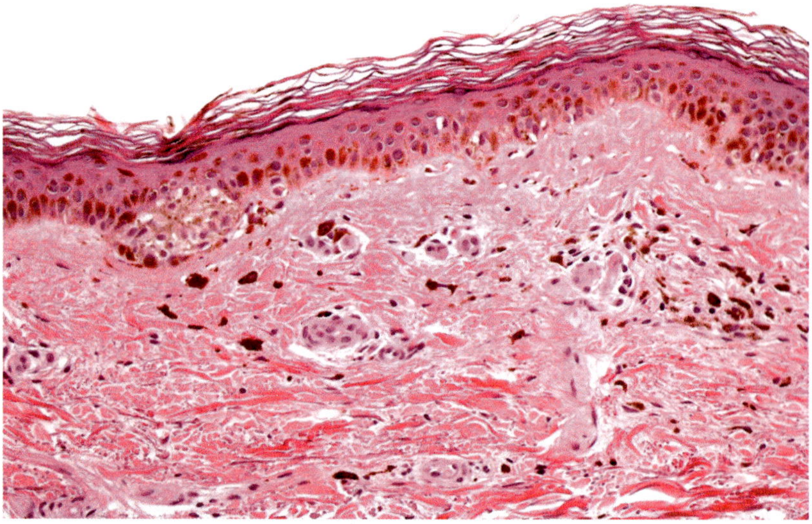

Fig. 1.5 The melanocytic proliferation was limited to the basal epidermal layer but it was continuous

Case 2

Patient: Male, 67 years old

Anatomic site of the lesion: Upper back

History of the lesion: Appeared several years earlier

Management: The dermoscopic findings were interpreted as suggestive of regression, and, therefore, the lesion was excised to rule out melanoma.

First histopathologic diagnosis: Junctional nevus

Second histopathologic diagnosis after clinicopathologic consultation: Early in situ melanoma

Key message: The well-known melanoma-specific dermoscopic criteria were investigated mainly in the context of invasive melanoma. Melanoma in situ usually does not display fully developed melanoma-specific criteria, but is characterized by only subtle dermoscopic features. The most frequent dermoscopic criteria of melanoma in situ are atypical network and regression. Consequently, a lesion displaying the latter features should be excised, especially in elderly patients in which reticular nevi are rarely seen.

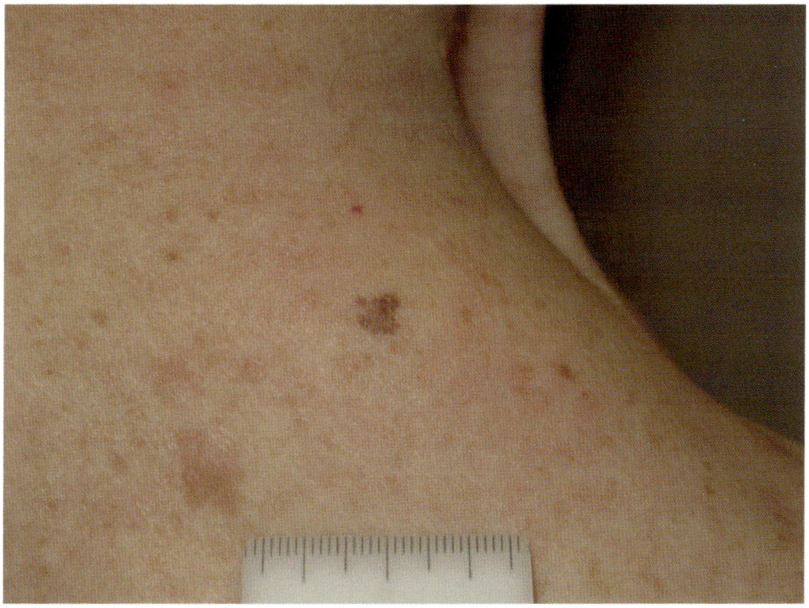

Fig. 1.6 A flat pigmented lesion on the upper back, characterized by light-brown color, relatively irregular border, and a maximum diameter of 6 mm

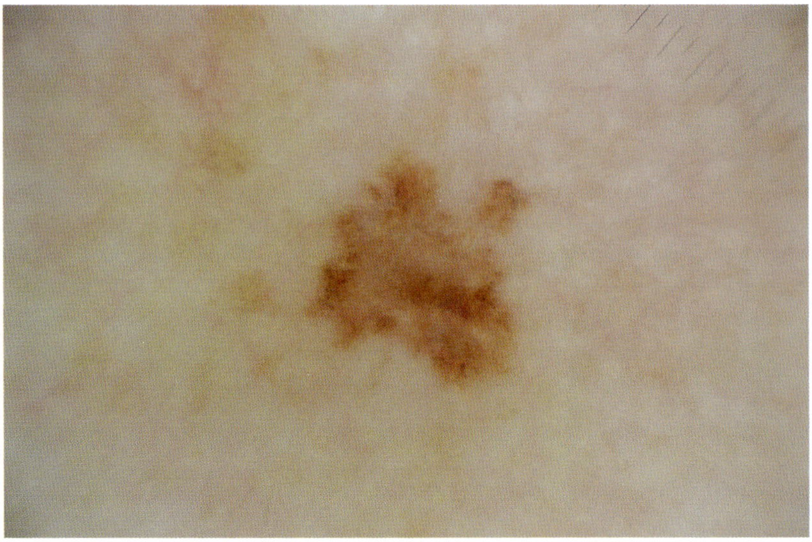

Fig. 1.7 Dermoscopically, the lesion displayed a reticular pattern. However, the network was not distributed symmetrically all over the surface of the lesion, a large part of which was hypopigmented. A few white lines and gray dots were also present

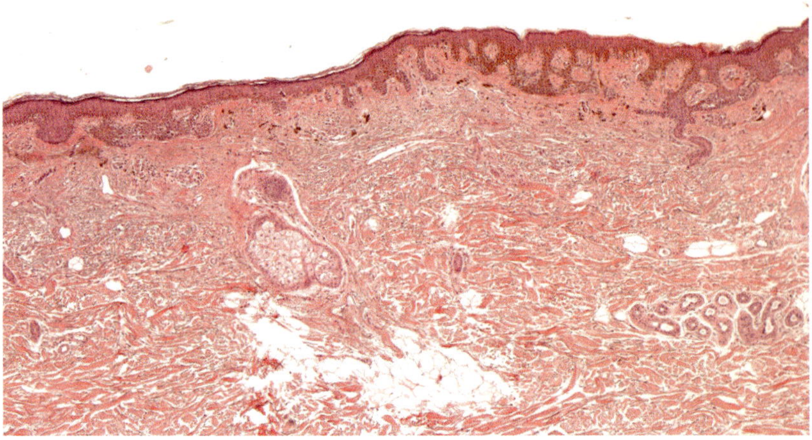

Fig. 1.8 Histologically, the lesion was constituted by a basally located, continuous proliferation of melanocytes. The epidermis was hyperplastic with fusion of the rete ridges

Figs. 1.9 and 1.10 The overall appearance was of a "dysplastic nevus." Some hyperchromatic atypical melanocytes were seen in the epidermis. In the dermis, fibrosis, some lymphocytes, and some melanophages were noted

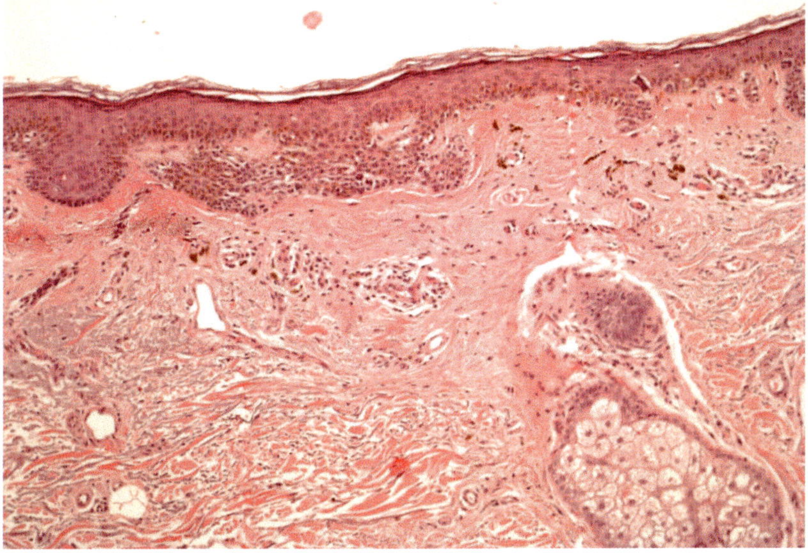

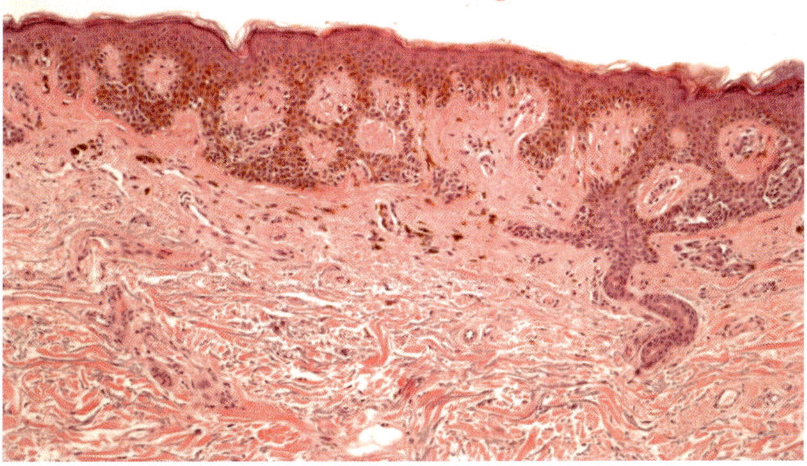

Case 3

Patient: Male, 76 years old

Anatomic site of the lesion: Middle back

History of the lesion: Unknown. A history of melanoma in situ on the trunk 20 years before

Management: The lesion was excised to rule out a melanoma.

Histologic diagnosis: Lentiginous melanoma in situ

Key message: Notably, the clinicopathologic correlation (in this case, age of the patient and dermoscopy) is crucial when histological features are subtle. The overall analysis of all morphologic aspects can help to avoid underdiagnosis.

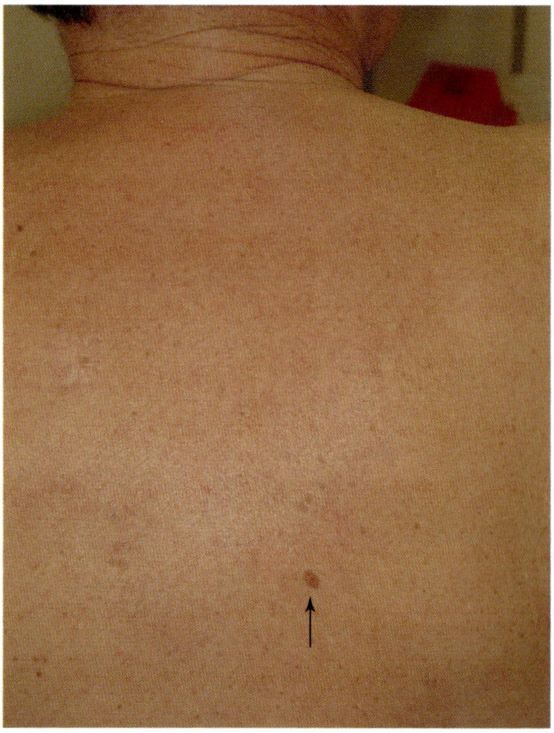

Fig. 1.11 Solitary lesion in the context of a patient without nevi

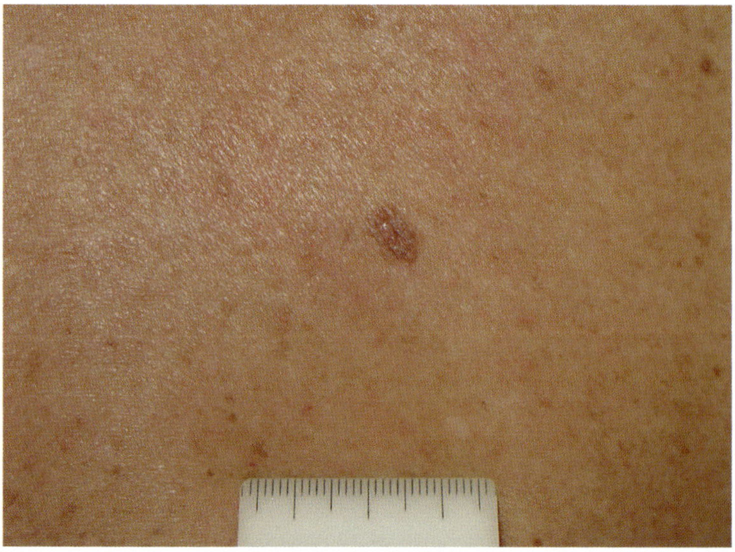

Fig. 1.12 Flat pigmented lesion with a maximum diameter of 8 mm

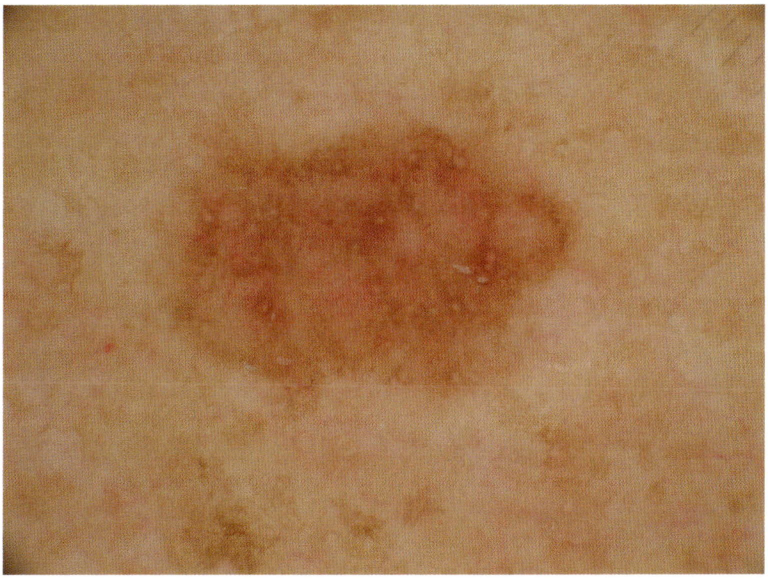

Fig. 1.13 Dermoscopically the lesion displayed an atypical pigmented network, focal inverse network, and barely visible gray color around the hair follicles

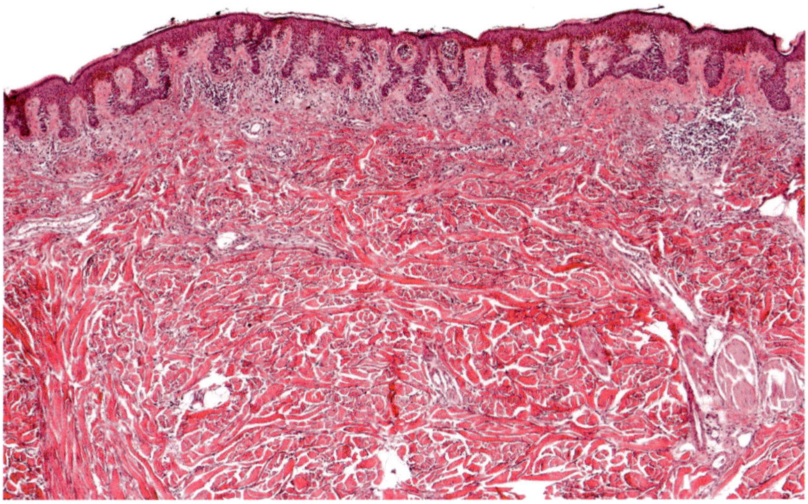

Fig. 1.14 On histology the lesion was relatively large, with ill-defined margins

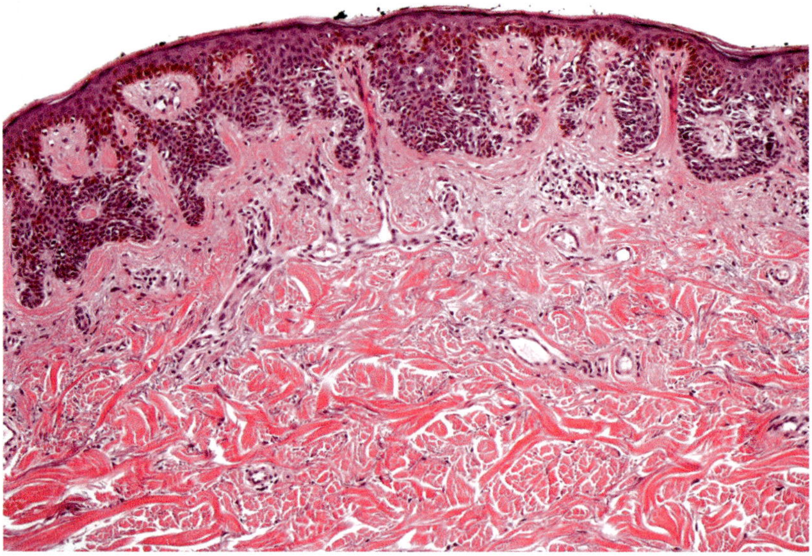

Fig. 1.15 At high-power view, there was a continuous, lentiginous melanocytic proliferation, basally located

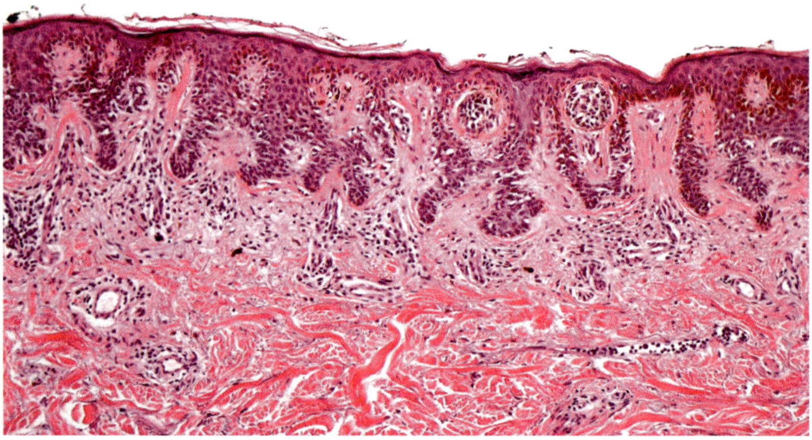

Fig. 1.16 Even if melanocytes were not strikingly atypical, this pattern of growth is more likely to be seen in melanoma in situ

Case 4

Patient: Female, 60 years old

Anatomic site of the lesion: Back

History of the lesion: Unknown

Management: The presence of regression, especially as it covered almost 50% of the lesion, prompted a surgical excision.

Histopathologic diagnosis: Compound nevus with lentiginous junctional hyperpigmented component

Key message: From a clinical point of view, the lesion was solitary and revealed a worrisome dermoscopic aspect with atypical hyperpigmented network and regression. These features are remarkably linked with melanoma diagnosis. However, in few cases, a worrisome dermoscopic aspect is not mirrored by a corresponding histopathologic diagnosis of melanoma, and we can explain the overall clinical morphology by the presence of hyperpigmentation that might be misleading.

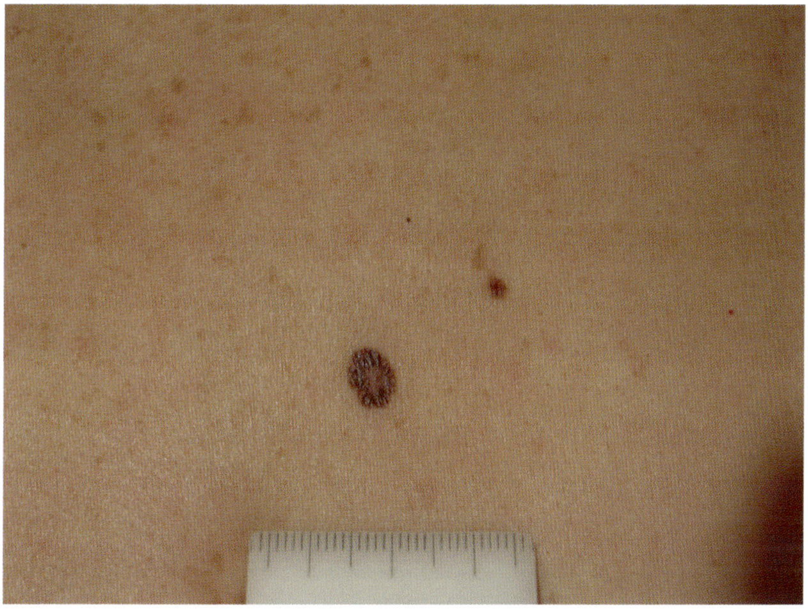

Fig. 1.17 Clinically the lesion was a dark-brown flat lesion with a diameter of 7 mm in this patient with a few nevi and solar lentigos

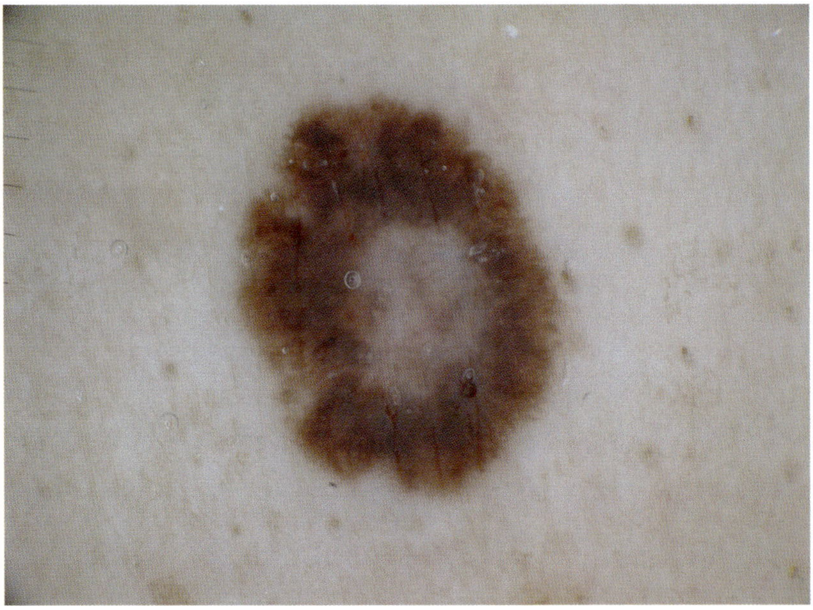

Fig. 1.18 Dermoscopically, the lesion showed an atypical pigment network at the periphery and a large central area of regression

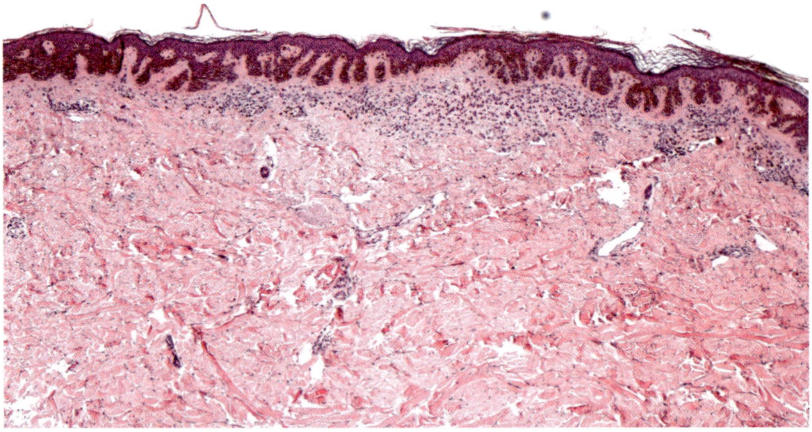

Fig. 1.19 Histologically, the lesion showed a junctional hyperpigmented component and some eccentric, small dermal melanocytes

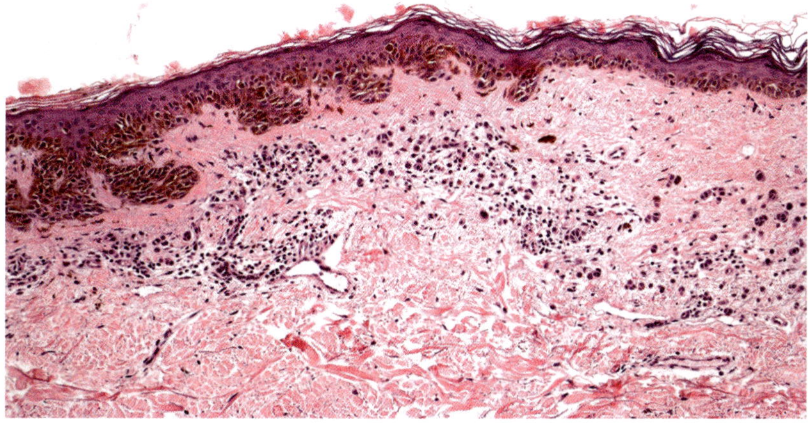

Fig. 1.20 The dermal melanocytes were cytologically bland, apparently mature, and with so-called ancient features

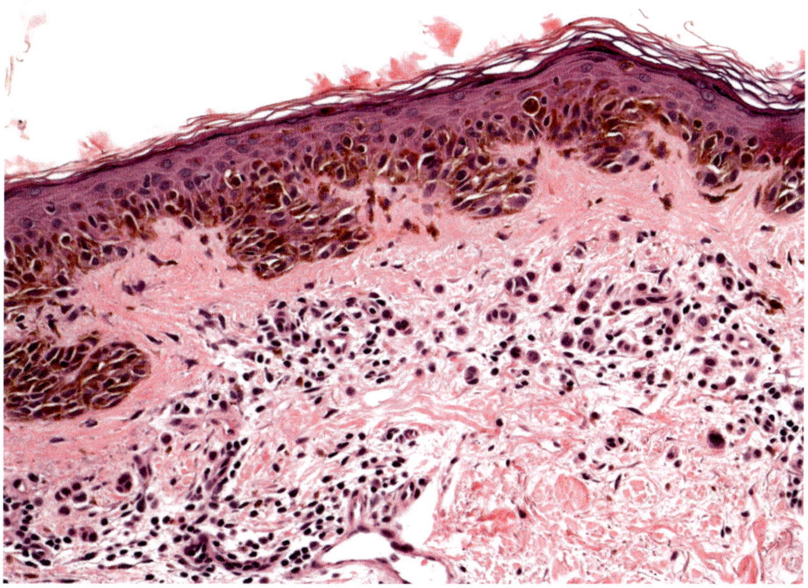

Fig. 1.21 At the junction, melanocytes were hyperpigmented and showed “random atypia.” Upward migration through the epidermis was absent

Case 5

Patient: Male, 75 years old

Anatomic site of the lesion: Back

History of the lesion: No relevant history

Management: The lesion was excised to rule out melanoma.

First histopathologic diagnosis: Junctional nevus

Second histopathologic diagnosis after clinicopathologic consultation: The critical analysis of the clinical and dermoscopic pictures prompted a revision of the original diagnosis. Due to the age of the patient and the actinic damage, the lesion was interpreted as an in situ melanoma.

Key message: A good clinicopathologic correlation in the management of junctional melanocytic proliferations in the elderly is extremely important. As the histologic diagnosis of early in situ melanoma can occasionally be difficult, clinicians must provide the dermatopathologist with essential clinical information and eventually critically review a pathologic report of a junctional or atypical lentiginous nevus, if it does not correlate with the clinical context of the patient.

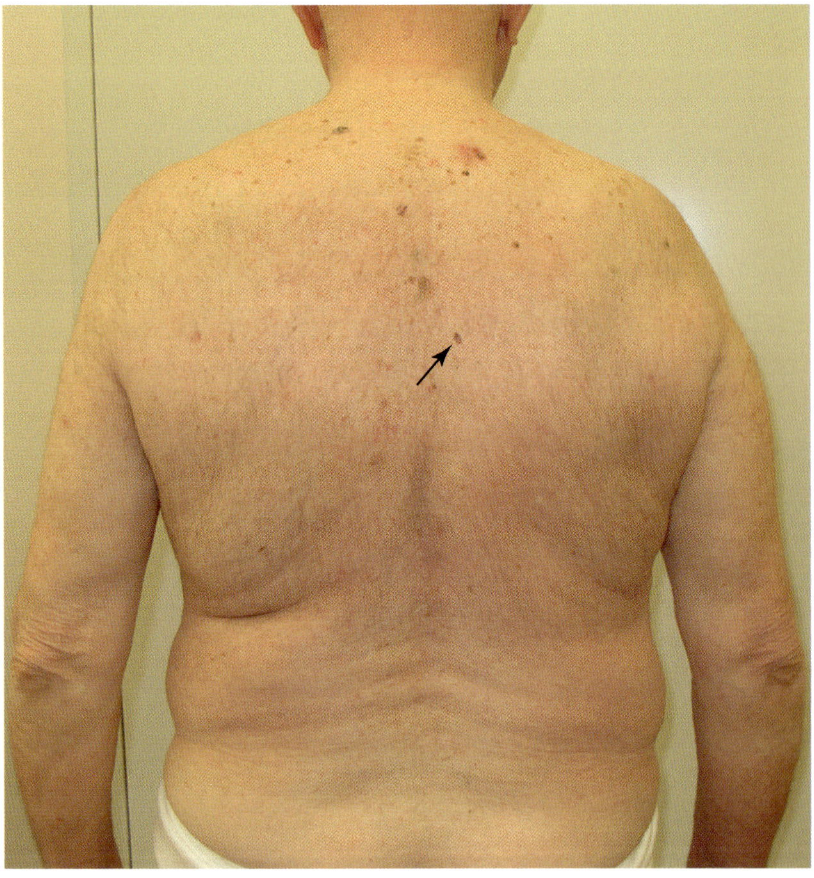

Fig. 1.22 The patient had moderate sun damage and multiple solar lentigos and seborrheic keratoses. Among them, there was a flat pigmented macule on the back

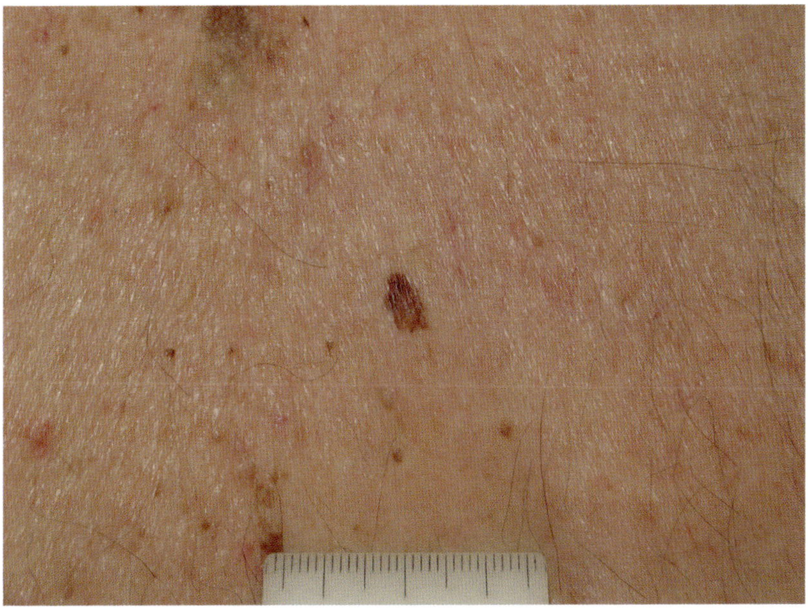

Fig. 1.23 Close-up clinical examination revealed a dark-brown lesion with a diameter of approximately 0.7 cm, with a slightly hypopigmented center and irregular borders

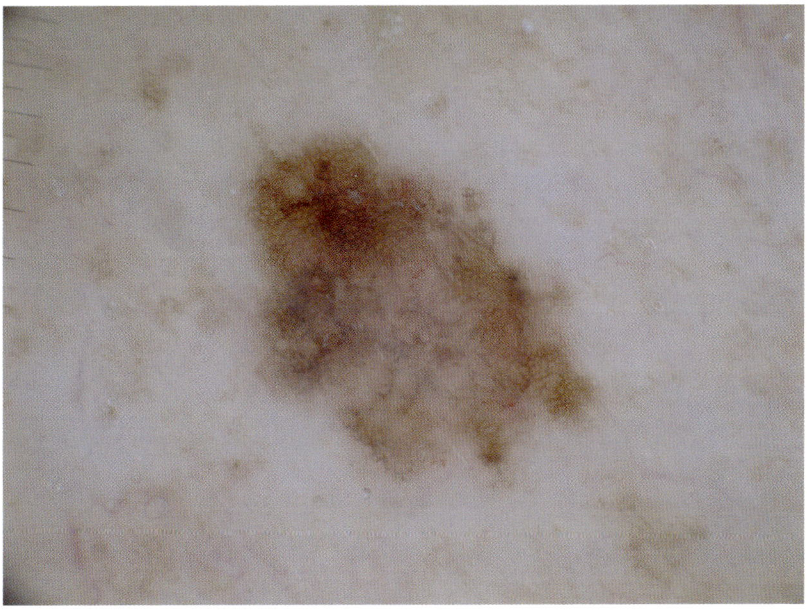

Fig. 1.24 Dermoscopically, the lesion was characterized by atypical network and regression structures. In the center of the lesion, fading of the dermoscopic structures was noticed, with multiple gray dots and globules

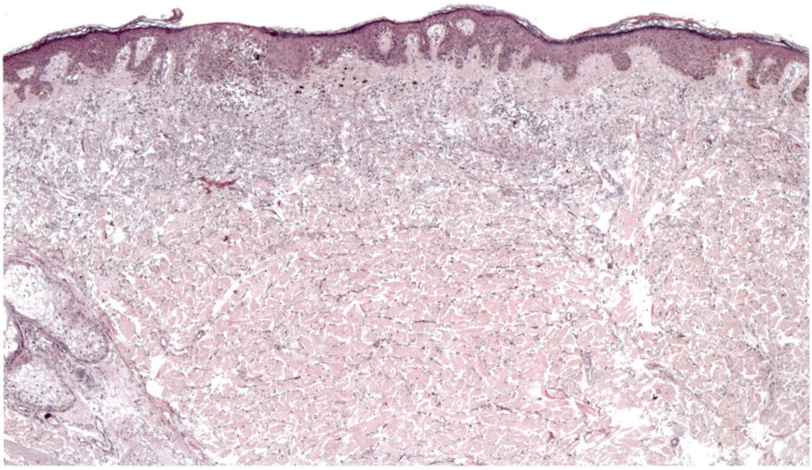

Fig. 1.25 On a background of severe sun-induced elastosis, a thin, continuous melanocytic proliferation with mild cytological atypia is present

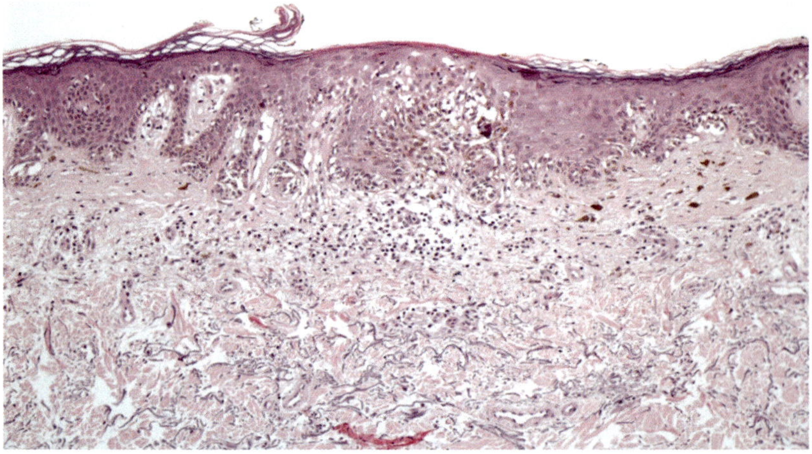

Fig. 1.26 Within a slightly hyperplastic epidermis, melanocytes were haphazardly juxtaposed

Suggested Reading

1. Argenziano G, Kittler H, Ferrara G, Rubegni P, Malvehy J, Puig S, Cowell L, Stanganelli I, De Giorgi V, Thomas L, Bahadoran P, Menzies SW, Piccolo D, Marghoob AA, Zalaudek I. Slow-growing melanoma: a dermoscopy follow-up study. Br J Dermatol. 2010;162(2):267–73.
2. Borsari S, Longo C, Piana S, Moscarella E, Lallas A, Alfano R, Argenziano G. When the 'Ugly Duckling' Loses Brothers, It Becomes the 'Only Son of a Widowed Mother'. Dermatology. 2015; 231(3):222–3.
3. Seidenari S, Bassoli S, Borsari S, et al. Variegated dermoscopy of in situ melanoma. Dermatology. 2012;224:262–70.
4. Argenziano G, Kittler H, Ferrara G, et al. Slow-growing melanoma: a dermoscopy follow-up study. Br J Dermatol. 2010;162:267–73.
5. Zalaudek I, Cota C, Ferrara G, Moscarella E, Guitera P, Longo C, Piana S, Argenziano G. Flat pigmented macules on sun-damaged skin of the head/neck: junctional nevus, atypical lentiginous nevus, or melanoma in situ? Clin Dermatol. 2014;32(1):88–93.
6. Massi D, De Giorgi V, Carli P, Santucci M. Diagnostic significance of the blue hue in dermoscopy of melanocytic lesions: a dermoscopic-pathologic study. Am J Dermatopathol. 2001;23(5):463–9.
7. Sardá SP, Suárez R, Malvehy Guilera J. Glowing in the dark: use of confocal microscopy in dark pigmented lesions. Dermatol Clin. 2016;34(4):431–42.
8. Okamura JM, Barr RJ, Cantos KA. Benign atypical junctional melanocytic hyperplasia associated with intradermal nevi: a common finding that may be confused with melanoma in situ. Mod Pathol. 2000;13(8):857–60.
9. Kerl H, Wolf IH, Kerl K, Ceroni L, Kutzner H, Argenyi ZB. Ancient melanocytic nevus. A simulator of malignant melanoma. Am J Dematopathol. 2011;33(2):127–30.
10. Ferrara G, Argenyi Z, Argenziano G, Cerio R, Cerroni L, Di Blasi A, Feudale EA, Giorgio CM, Massone C, Nappi O, Tomasini C, Urso C, Zalaudek I, Kittler H, Soyer HP. The influence of clinical information in the histopathologic diagnosis of melanocytic skin neoplasms. PLoS One. 2009;4(4):e5375.
11. Longo C, Piana S, Lallas A, Moscarella E, Lombardi M, Raucci M, Pellacani G, Argenziano G. Routine clinical-pathologic correlation of pigmented skin tumors can influence patient management. PLoS One. 2015;10(9):e0136031.
12. Ferrara G, Annessi G, Argenyi Z, Argenziano G, Beltraminelli H, Cerio R, Cerroni L, Cota C, Simonetti S, Stefanato CM, Zalaudek I, Kittler H, Soyer HP. Prior knowledge of the clinical picture does not introduce bias in the histopathologic diagnosis of melanocytic skin lesions. J Cutan Pathol. 2015;42(12):953–8.

2 Melanocytic Atypical Lesions in Patients with Multiple Nevi

In patients with multiple nevi, management goals include early diagnosis of melanoma while avoiding unnecessary excisions of benign lesions.

Dealing with patients with multiple melanocytic lesions, the clinical information should include the patient's age, skin photo-type, body location, changes, and genetic background.

In the physical examination of patient lesions, a comparative analysis approach is advocated to detect an *ugly duckling* lesion. According to this strategy, melanocytic lesions in patients tend to follow a predominant pattern of colors and size. The *ugly duckling* lesion differs from other patient lesions. This strategy has been postulated as more efficient for the detection of melanoma in patients with multiple nevi with respect to the analytical approach of the ABCD (asymmetry, border irregularity, color variegation, diameter greater than 6 mm) rule.

However, the *ugly duckling* sign is not always sufficient to discriminate an early melanoma in the context of patients with many atypical looking nevi. In fact, melanoma in these patients may be similar to other lesions, and the correlation of the clinical atypia with histopathologic atypia or even melanoma is extremely low.

In patients with multiple atypical nevi, early melanomas cannot always be distinguished with accuracy, even with the use of dermoscopy by experienced dermatologists.

For such cases, a dermoscopic comparative approach was postulated. It is well-known that one or two predominant dermoscopic patterns can be found in patients with multiple nevi. According to this concept, the key point in the examination of individuals with multiple nevi is the identification of their predominant nevus pattern (defined as the pattern observed in more than 30% of all melanocytic lesions), which then allows the recognition of atypical lesions that deviate from this pattern.

An additional consideration in patients with high risk for melanoma is that the predominant nevus pattern of patients with melanoma seems to differ from that of the healthy population.

In recent studies, a dermoscopic complex or multicomponent pattern (defined in melanocytic lesions exhibiting both network and globules, with or without structureless areas) was found significantly more frequently in melanoma patients than in controls, who usually have nevi with uniform dermoscopic patterns (i.e., reticular pattern). As a general rule, individuals with multiple nevi may benefit from closer surveillance with dermoscopic digital monitoring compared to individuals with few nevi or with melanocytic lesions exhibiting uniform patterns.

C. Longo et al., *Atlas of Diagnostically Challenging Melanocytic Neoplasms*,
https://doi.org/10.1007/978-3-319-48653-6_2

Case 1

Patient: Female, 39 years old

Anatomic site of the lesion: Left thigh

History of the lesion: Appeared several years ago and gradually enlarged

Management: The lesion was excised to rule out melanoma.

Histopathologic diagnosis: Compound nevus

Key message: It is well-known that several nevi of patients with the so-called atypical mole syndrome might display severe clinical and dermoscopic atypia. Management of such patients requires a careful clinical and dermoscopic examination and a morphologic comparison among the moles in order to detect the "ugly duckling," namely, the mole looking clinically or dermoscopically different. Although this approach significantly reduces the number of unnecessary excisions of nevi, still some nevi will be excised, since their morphologic atypia extends beyond the tolerable threshold. The histopathologic diagnosis of a nevus is acceptable in the context of relatively young patients with many atypical moles, even if their clinical and dermoscopic morphology was worrisome.

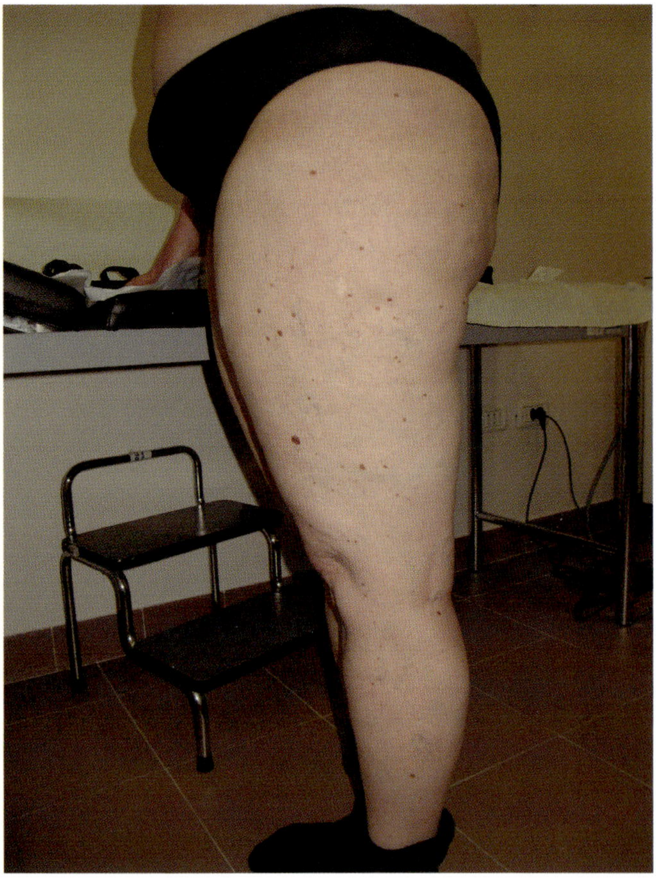

Fig. 2.1 The patient had several moles on the trunk and the extremities. Among them, there was a relatively large mole on the left thigh

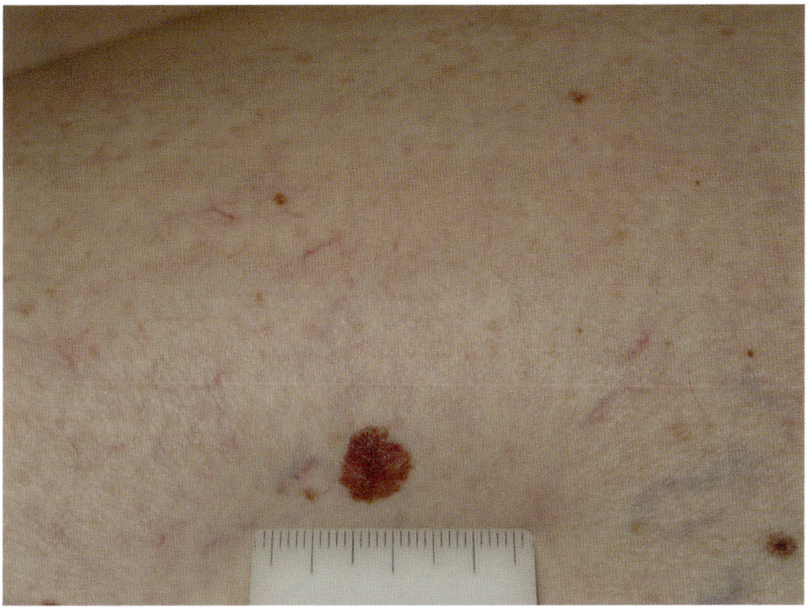

Fig. 2.2 Close-up clinical examination revealed a roundish brown lesion with a diameter of approximately 1 cm, with a slightly hyperpigmented center and an irregular border

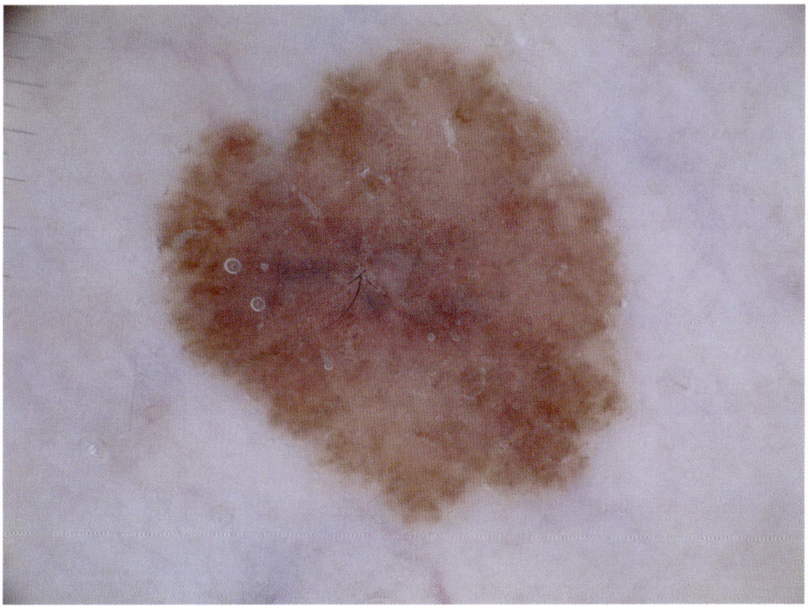

Fig. 2.3 Dermoscopically, the lesion was characterized by uneven distribution of pigmentation, irregularly arranged brown globules, a few blue-gray globules in the center, and tan structureless areas at the periphery

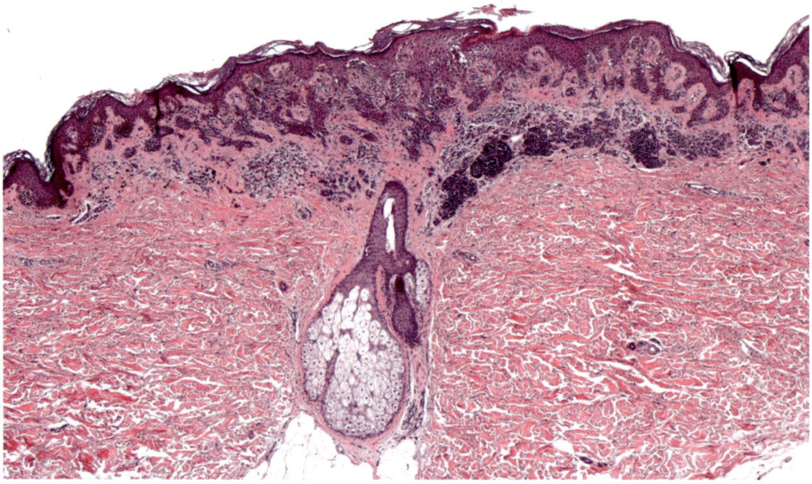

Fig. 2.4 Histologically, the lesion showed an evident junctional component, with some spitzoid features and some mature-looking, congenital-type dermal nests

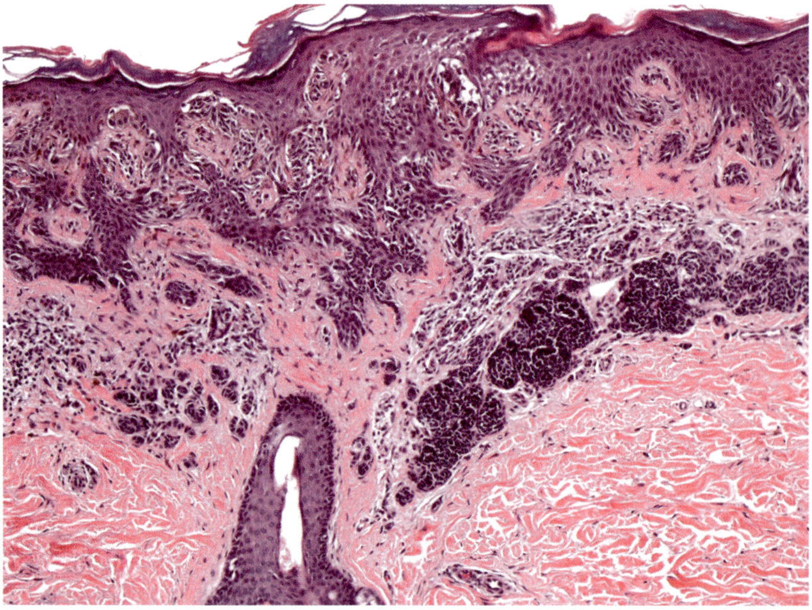

Fig. 2.5 Even if prominent, the junctional melanocytes did not show upward epidermal migration

Case 2

Patient: Female, 19 years old

Anatomic site of the lesion: Back

History of the lesion: Present since many years, recently changed in pigmentation. Many moles are present on the trunk.

Management: The lesion was excised to rule out melanoma arising on congenital nevus.

Histopathologic diagnosis: Congenital-type compound nevus

Key message: The detection of a clinically nodular area arising in the context of a long-standing nevus should raise the suspicion of a melanoma arising on congenital nevus. In dermoscopy, congenital nevi are characterized by cobblestone pattern, whereas blue-white veil is a well-known melanoma feature. Histopathologic examination is mandatory in these cases to rule out melanoma.

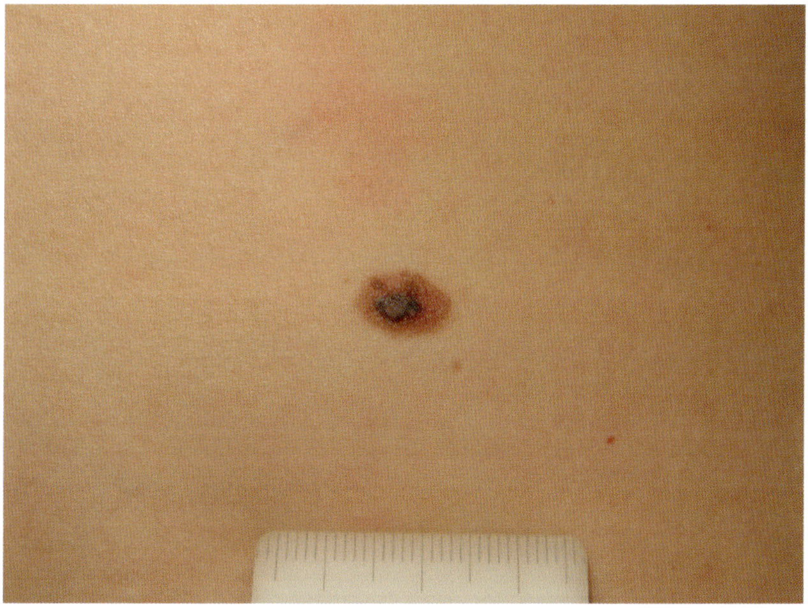

Fig. 2.6 Close-up clinical examination revealed a roundish pigmented mole, with a palpable, blue to gray area in the center of the lesion

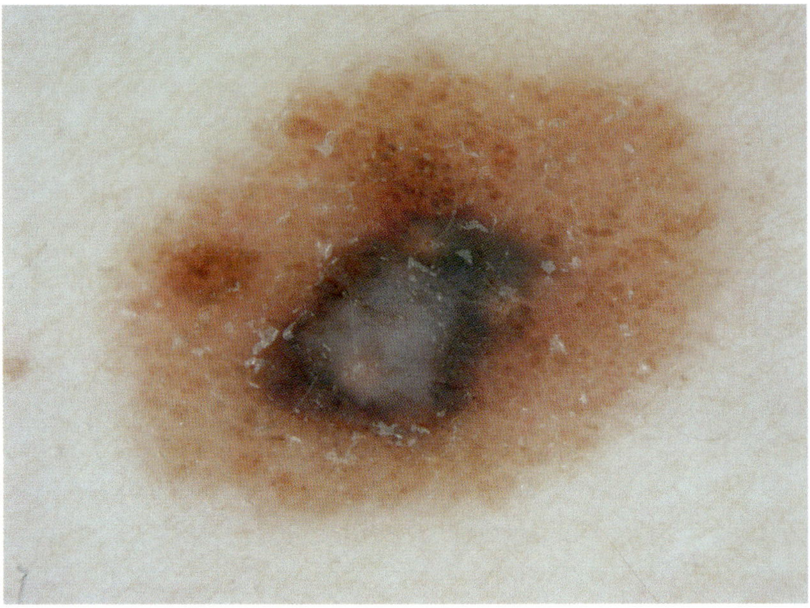

Fig. 2.7 Dermoscopically, the lesion was characterized by an overall regular globular pattern, interrupted in the center by a blue-white veil

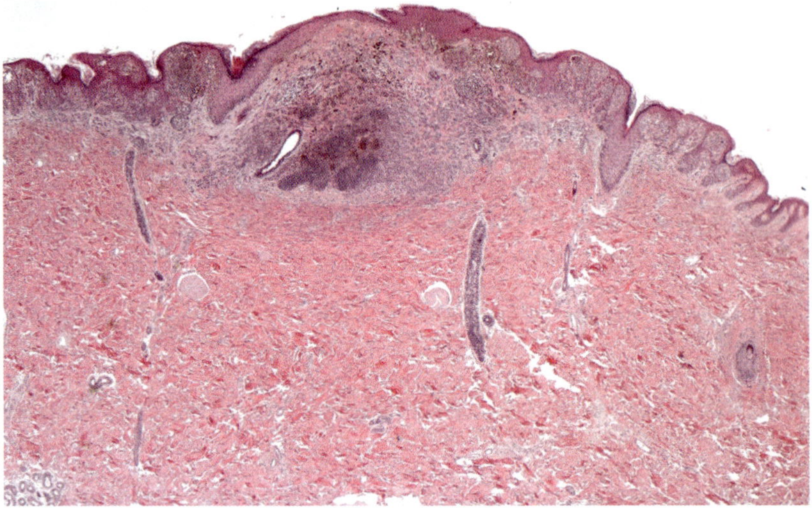

Fig. 2.8 The lesion was asymmetric and showed a dermal nodular component with large proliferating "shoulders"

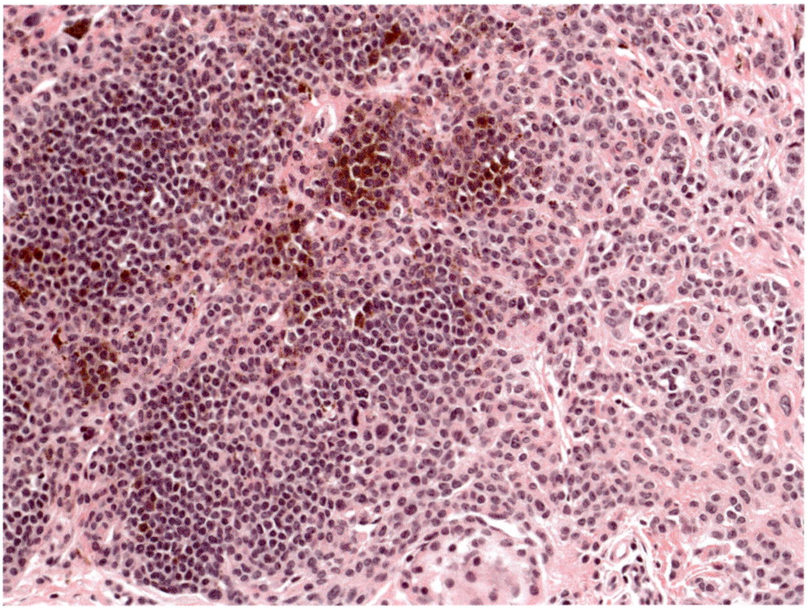

Fig. 2.9 The dermal nodule was made up of densely packed usual melanocytes, devoid of mitotic activity, and it was reminiscent of the so-called proliferative nodule

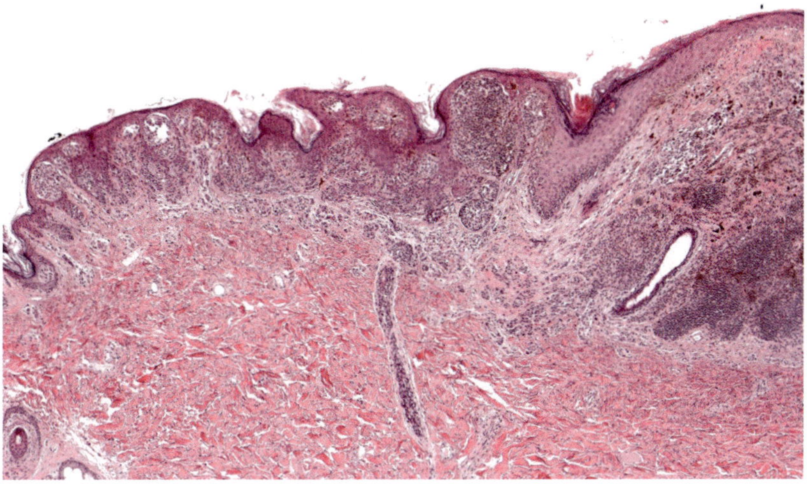

Fig. 2.10 In the shoulders, melanocytes were arranged in large nests without epidermotropism

Case 3

Patient: Male, 62 years old

Anatomic site of the lesion: Right flank

History of the lesion: The patient had several moles on the trunk and extremities. Among them, was a solitary lesion.

Management: The lesion was excised to rule out melanoma although the differential diagnosis included a congenital nevus.

Histopathologic diagnosis: Melanoma in situ arising on dermal nevus

Key message: Typically, a congenital nevus might display a regular cobblestone pattern and a history of a long-standing lesion. In the present case, the presence of asymmetric distribution of color and structures raised the suspicion that the lesion could be a nevus-associated melanoma. The distinction between nevus component and melanoma was extremely difficult since the malignant proliferation was located on top of the banal dermal nevus. In general, the presence of inverse network is likely to be seen in nevus-associated melanoma although in our case this criterion was missing.

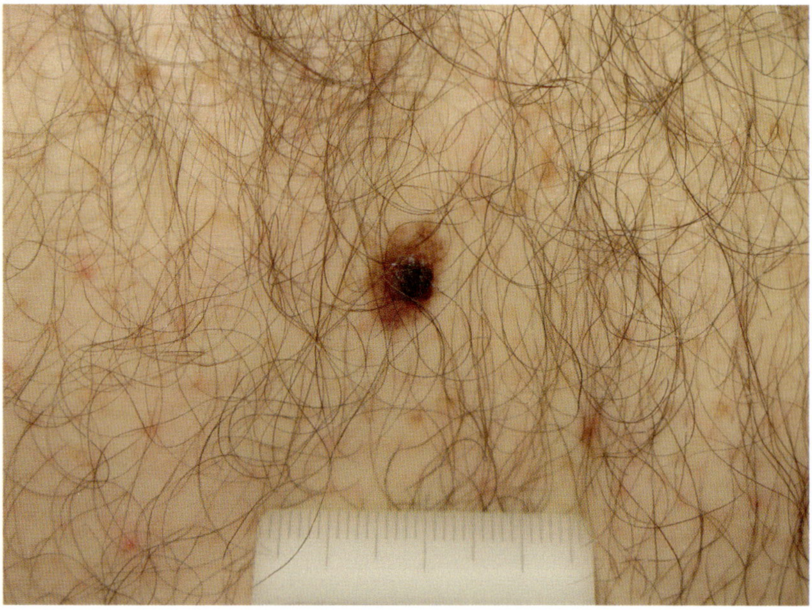

Fig. 2.11 Close-up clinical examination revealed an asymmetric large lesion pigmented lesion with a diameter of 1 × 0.5 cm, with irregular borders and central hyperpigmentation

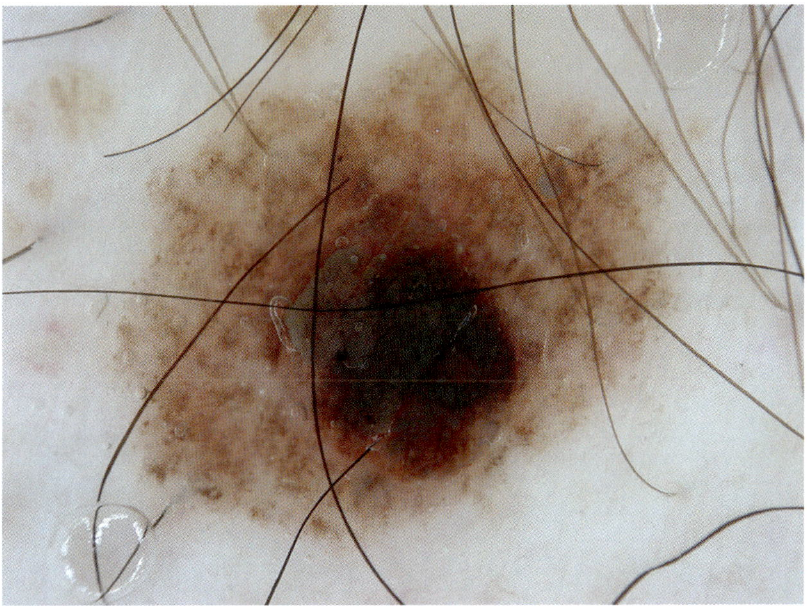

Fig. 2.12 Dermoscopically, the lesion was characterized by a light-brown flat portion with reticular pattern and few globules and a central dark-brown area with papillomatous surface

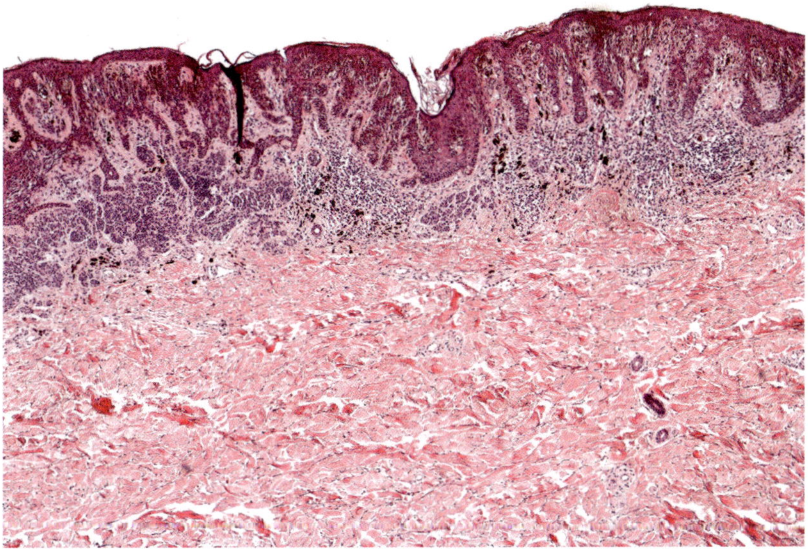

Fig. 2.13 At scanning magnification, the lesion was asymmetrical with an evident junctional component. On the left, some dermal, benign-looking, melanocytic nests

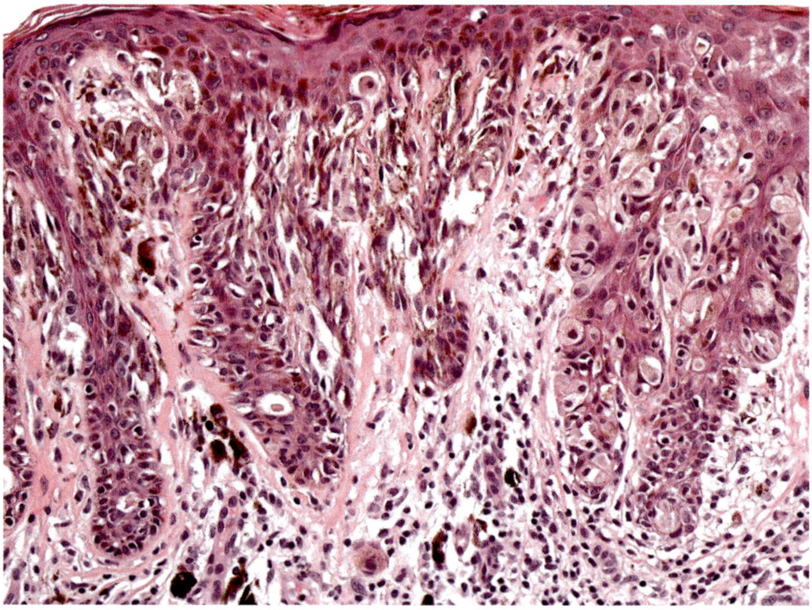

Fig. 2.14 At high-power view, the junctional proliferation was atypical, composed of pleomorphic, epithelioid melanocytes, limited to the epidermal surface

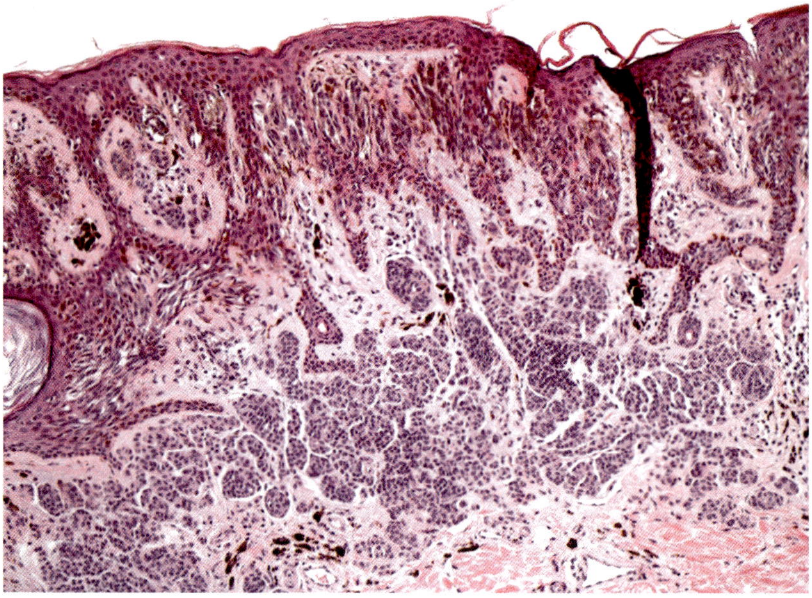

Fig. 2.15 Dermal nests were eccentrically located, monomorphous, and easily distinguishable from the superficial atypical melanocytes

Case 4

Patient: Female, 44 years old

Anatomic site of the lesion: Upper back

History of the lesion: The patient had severely sun-damaged skin on the trunk and the extremities. A relatively large mole on the right scapula was noticed among multiple solar lentigos.

Management: The lesion was excised to rule out melanoma.

First histopathologic diagnosis: Junctional nevus

Second histopathologic diagnosis after clinicopathologic consultation: The diagnosis was revised after considering the clinical and dermoscopic features. Although histological features were very subtle, a very early melanoma in situ with some regression could not be ruled out.

Key message: In the field of melanocytic lesions, there are cases for which definitive criteria simply do not exist, and therefore there is a worrisome interpersonal variability, even among experts. In equivocal melanocytic neoplasms, both the thin lesions and the nodular ones, the dermatopathological evaluation carries a high risk of over- or underdiagnosis, thus causing potentially diagnostic errors. The role of the clinicians is to contribute through expert clinical assessment and clear communication with the dermatopathologist. In addition, the dermatopathologist should be cautious in the reporting, being aware of the abovementioned variability and asking for safe margins whenever necessary.

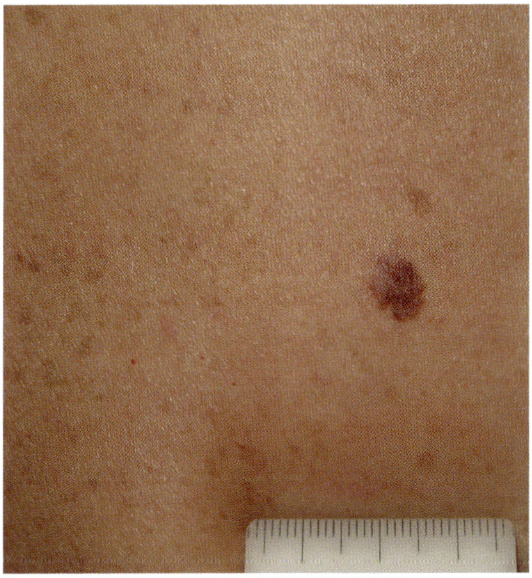

Fig. 2.16 Close-up clinical examination revealed a roundish brown lesion with a diameter of 0.9 cm, with uneven distribution of pigmentation and irregular borders

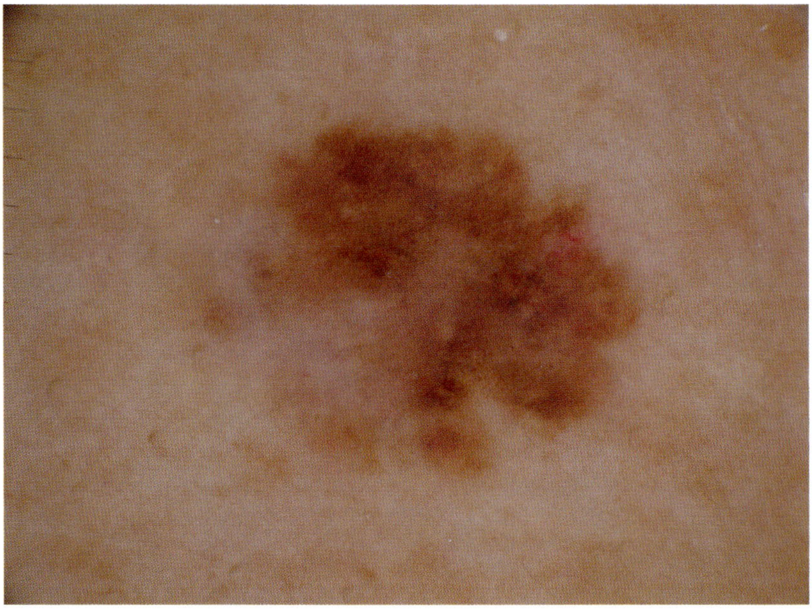

Fig. 2.17 Dermoscopically, the lesion was highly asymmetric, characterized by atypical pigment network and regression structures. White scar-like areas and gray globules involved asymmetrically about 50% of lesion surface

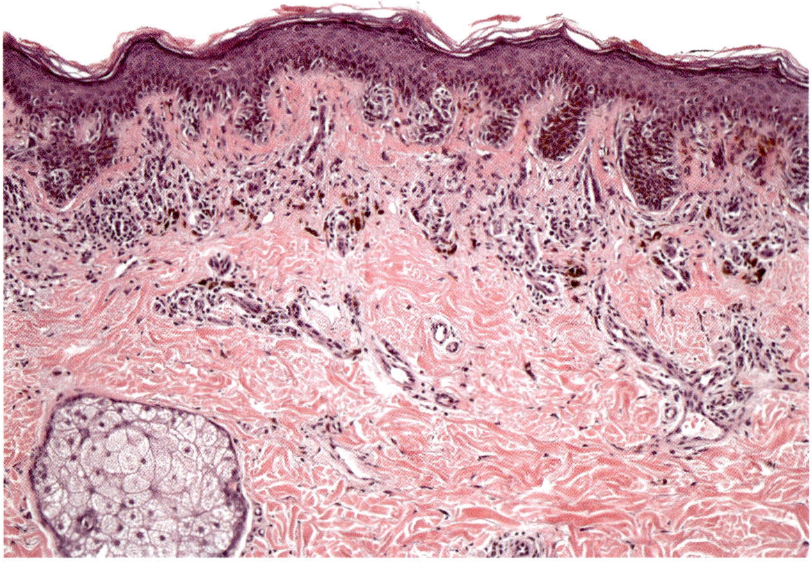

Fig. 2.18 The lesion was junctional and melanocytes were arranged in small nests and single units

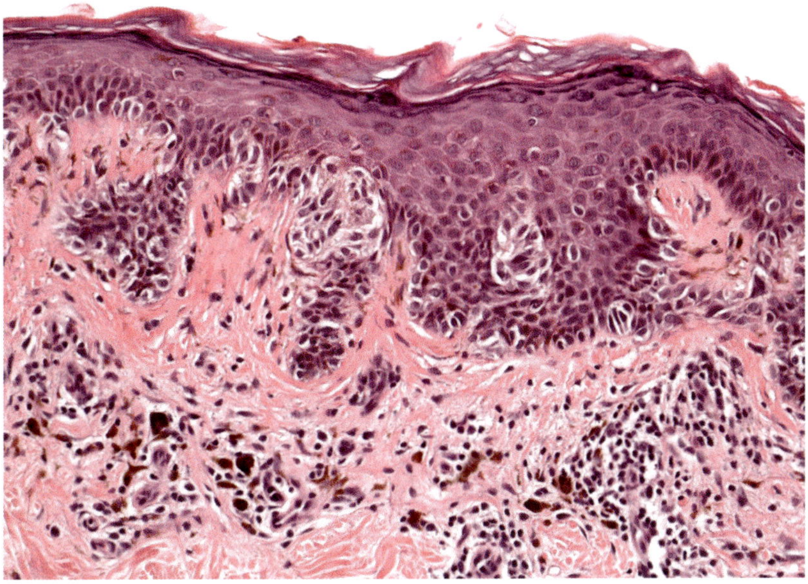

Fig. 2.19 Cellular pleomorphism was minimal and upward migration through the epidermis was absent. In the dermis, there were a mild inflammatory infiltrate and many melanophages

Case 5

Patient: Female, 42 years old

Anatomic site of the lesion: Upper limb

History of the lesion: The patient had several moles on the trunk and the extremities. Among them, there was a relatively large mole on the left arm, appeared several years ago and gradually enlarged.

Management: The lesion was excised to rule out melanoma.

Histopathologic diagnosis: Compound nevus

Key message: It is well-known that several nevi of patients with the so-called atypical mole syndrome might display severe clinical and dermoscopic atypia. Management of such patients requires a careful clinical and dermoscopic examination and a morphologic comparison among the moles in order to detect the "ugly duckling," namely, the nevus looking clinically or dermoscopically different. The histopathologic diagnosis of a nevus is acceptable in the context of relatively young patients with many atypical moles, even if their clinical and dermoscopic morphology was worrisome.

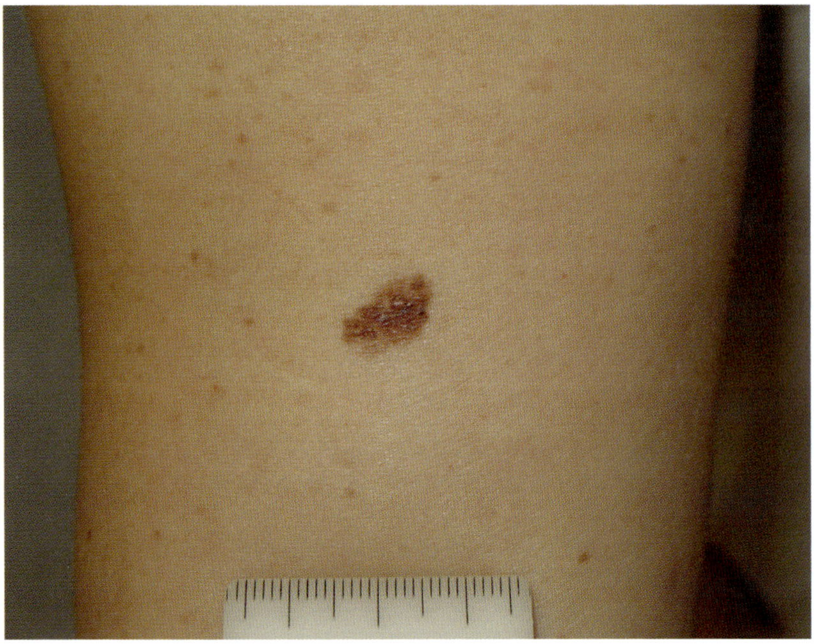

Fig. 2.20 Close-up clinical examination revealed a brown macule with a diameter of approximately 1.3 cm, with uneven pigmentation and irregular borders

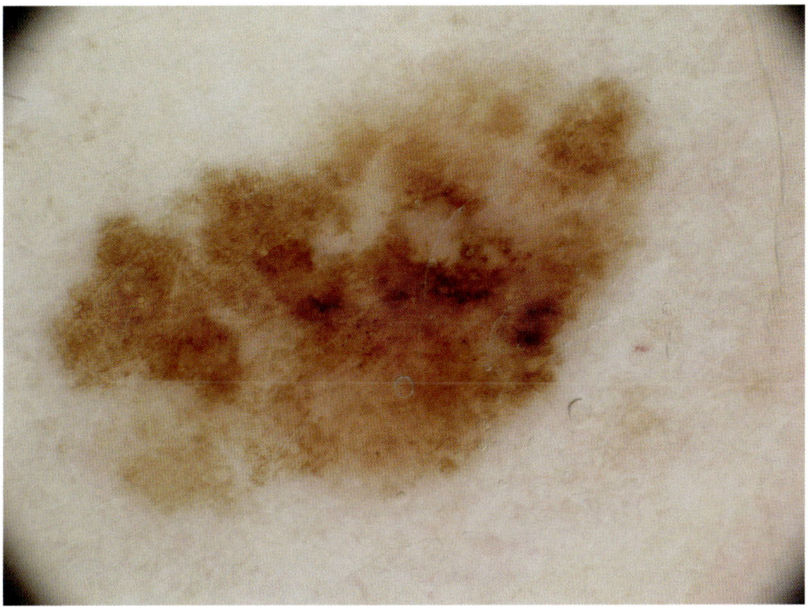

Fig. 2.21 Dermoscopically, the lesion was highly asymmetric, with irregular pigment network and multiple irregular brown to black globules in the center of the lesion. Structureless brown pigmentation and fading of dermoscopic structures were seen at the periphery of the lesion

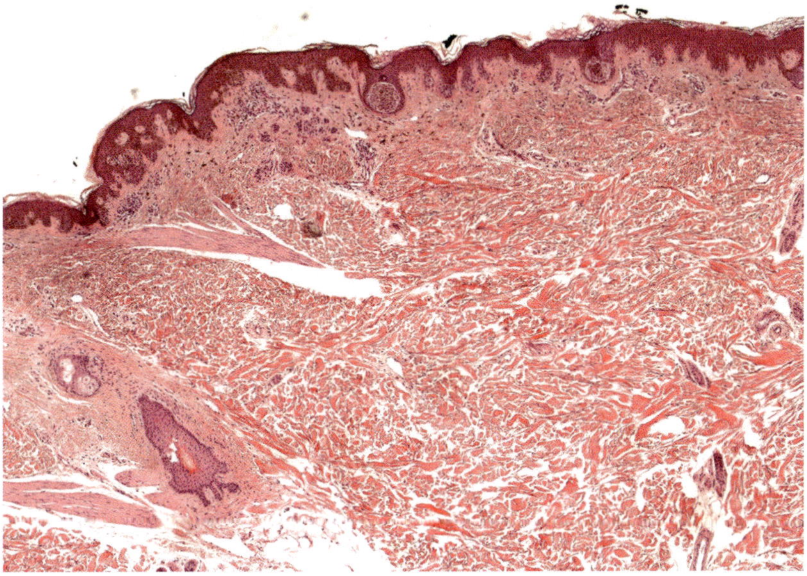

Fig. 2.22 The lesion was asymmetrical and scarcely populated; occasional melanocytes at the junction and some dermal nests at the border were visible

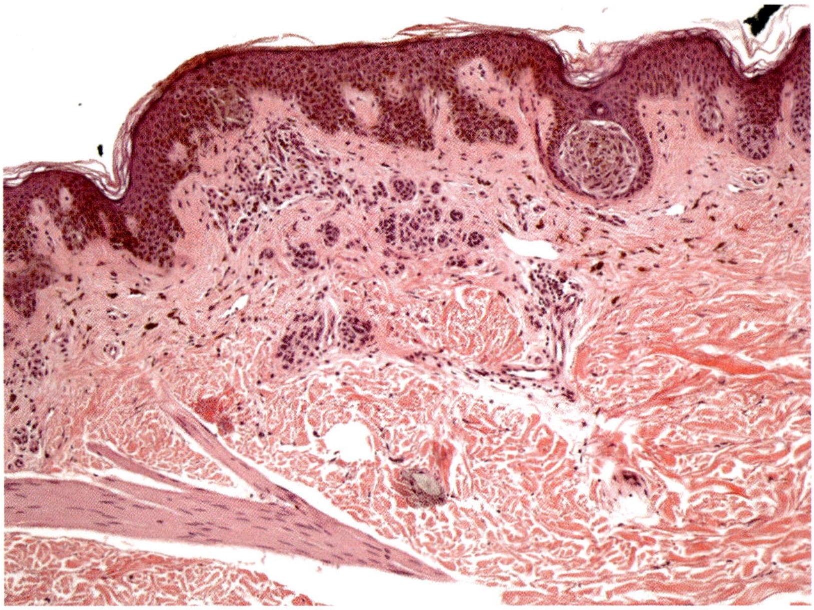

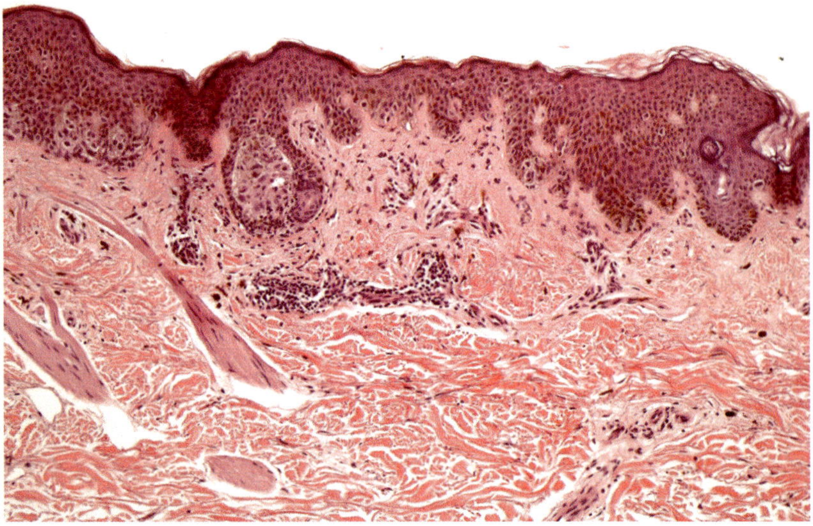

Figs. 2.23 and 2.24 Some junctional melanocytic nests overlay a benign-appearing, peripheral dermal component

Suggested Reading

1. Argenziano G, Catricalà C, Ardigo M, et al. Dermoscopy of patients with multiple nevi: improved management recommendations using a comparative diagnostic approach. Arch Dermatol. 2011;147:46–9.
2. Lallas A, Zalaudek I, Apalla Z, et al. Management rules to detect melanoma. Dermatology. 2013;226:52–60.
3. Massi G. Melanocytic nevi simulant of melanoma with medicolegal relevance. Virchows Arch. 2007;451(3):623–47.
4. Yélamos O, Arva NC, Obregon R, Yazdan P, Wagner A, Guitart J, Gerami P. A comparative study of proliferative nodules and lethal melanomas in congenital nevi from children. Am J Surg Pathol. 2015;39(3):405–15.
5. Haenssle HA, Mograby N, Ngassa A, Buhl T, Emmert S, Schön MP, Rosenberger A, Bertsch HP. Association of patient risk factors and frequency of nevus-associated cutaneous melanomas. JAMA Dermatol. 2016;152(3):291–8.
6. Shitara D, Nascimento M, Ishioka P, Carrera C, Alós L, Malvehy J, Puig S. Dermoscopy of Naevus-associated melanomas. Acta Derm Venereol. 2015;95(6):671–5.
7. Longo C, Piana S, Lallas A, Moscarella E, Lombardi M, Raucci M, Pellacani G, Argenziano G. Routine clinical-pathologic correlation of pigmented skin tumors can influence patient management. PLoS One. 2015;10(9):e0136031.
8. Scope A, Dusza SW, Halpern AC, Rabinovitz H, Braun RP, Zalaudek I, Argenziano G, Marghoob AA. The "ugly duckling" sign: agreement between observers. Arch Dermatol. 2008;144(1):58–64.

3 Lesions on the Head and Neck

Head and neck area represents a "special" body site where the close collaboration of the dermatologist together with the pathologist is of uppermost importance for reaching a correct diagnosis and consequently an optimal patient management. On this body site, especially on chronically sun-damaged skin in adults and elderly individuals, clinical and dermoscopic differential diagnosis can be challenging.

Clinical and dermoscopic differential diagnosis includes benign pigmented macules (solar lentigo, actinic keratosis, lichen planus-like keratosis) to be differentiated from lentigo maligna (LM). The dermatoscopic aspects of melanocytic lesions on this special body sites differ from the ones described for other locations. A pigment network, representing the dermatoscopic hallmark of melanocytic tumors on the torso and extremities, is only rarely detected. This is because pigment network results from epidermal melanin along elongated rete ridges, whereby the tips of the rete ridges appear as network holes and the lateral borders as network lines. Instead, the dermoepidermal junction of chronically sun-damaged facial skin appears flattened and may even lack rete ridges. Pigmented keratinocytes or melanocytes along this flattened dermoepidermal junction appear as structureless diffuse brown pigmentation on dermoscopy. This structureless diffuse brown pigmentation is interrupted by numerous, variably broad, and hypopigmented "holes," which correspond to hair follicles and sweat gland openings, and it is called "pseudonetwork" pattern. The pseudonetwork pattern occurs likewise in melanocytic and non-melanocytic lesions. This is why the differential diagnosis relays on additional dermoscopic clues, such as the detection of gray color, especially when arranged in circles or in the so-called rhomboidal structures.

Moreover, it has to be underlined that also the histopathologic diagnosis can be challenging because early LM can display very subtle features on histopathology and the lesion can be overlooked, especially when clinical information is absent. In order to minimize the risk to underdiagnosing a melanoma as a junctional nevus, clinicopathologic correlation is mandatory. Clinicians should always pay particular attention if confronted with a given histopathologic diagnosis of a "junctional" or "dysplastic nevus" on the head/neck area of an elderly patient and critically review such diagnosis in the context of the patient considering the history, age, sex, location, and clinical or dermatoscopic features of the lesion.

In addition, when dealing with a large and flat pigmented lesion, a partial biopsy is a common practice. This is because large excisions might be disfigurant, and in case of clinically doubtful lesions, a partial biopsy is sometimes preferred. A shave biopsy can adequately sample the entire lesion, but if an incision biopsy is preferred, the latter must be aimed at differentiating a melanocytic from a non-melanocytic proliferation, i.e., lentigo maligna (melanoma) from solar

C. Longo et al., *Atlas of Diagnostically Challenging Melanocytic Neoplasms*,
https://doi.org/10.1007/978-3-319-48653-6_3

lentigo and pigmented actinic keratosis. If the lesions are histopathologically proven to be melanocytic on an intralesional biopsy, this is enough to prompt in toto excision, because a "melanocytic nevus" is not in the range of the clinical differential diagnoses. A melanocytic hyperplasia on chronically sun-exposed skin can be also advocated in the differential diagnosis, but it is excluded when melanocytic nests are present at histological examination. On the other hand, pathologists must be cautious when dealing with an incisional biopsy of a melanocytic proliferation in sun-exposed sites as it could not be representative of the entire lesion.

This section shows some clinical, dermoscopic, and histopathologic features of problematic skin tumors on the head/neck area.

Case 1

Patient: Male, 76 years old

Anatomic site of the lesion: Forehead

History of the lesion: Since several years. The patient referred that cryotherapy was performed in the past.

Management: The clinical and dermoscopic image were not highly specific. The differential diagnosis included lentigo maligna, seborrheic keratosis, and pigmented actinic keratosis. The gray dots/granules possibly corresponded to melanophages, which can be seen in all the three entities mentioned. The light-brown-colored circles and parallel lines were rather suggestive of solar lentigo/seborrheic keratosis. However, the presence of dark-brown dots raised the suspicion of lentigo maligna. A dermoscopy-guided punch biopsy was performed from the part of the lesion displaying dark-brown dots.

Histopathologic diagnosis: In situ melanoma in sun-damaged skin

Key message: The diagnostic approach of pigmented facial macules is very challenging, because the entities included in the differential diagnosis (lentigo maligna melanoma, pigmented actinic keratosis, and solar lentigo/seborrheic keratosis) share similar clinical and dermoscopic characteristics. Gray color has been suggested to represent the most frequent dermoscopic finding of lentigo maligna melanoma. However, the specificity of gray color is low, since it can also be seen in the majority of pigmented actinic keratoses, as well as in a solar lentigo/seborrheic keratosis undergoing regression (lichen planus-like keratosis). The presence of pigment at the edge of the hair follicles possibly represents the most useful clue of lentigo maligna melanoma, since in the other two entities the pigmentation typically tends to respect the follicular openings. Punch biopsies might be appropriate in facial lesions, as the differential diagnosis lays between melanoma and non-melanocytic tumors. However, punch biopsies should always be dermoscopy-guided, since the possibility of collisions between lentigo maligna and solar lentigo or actinic keratosis is not low.

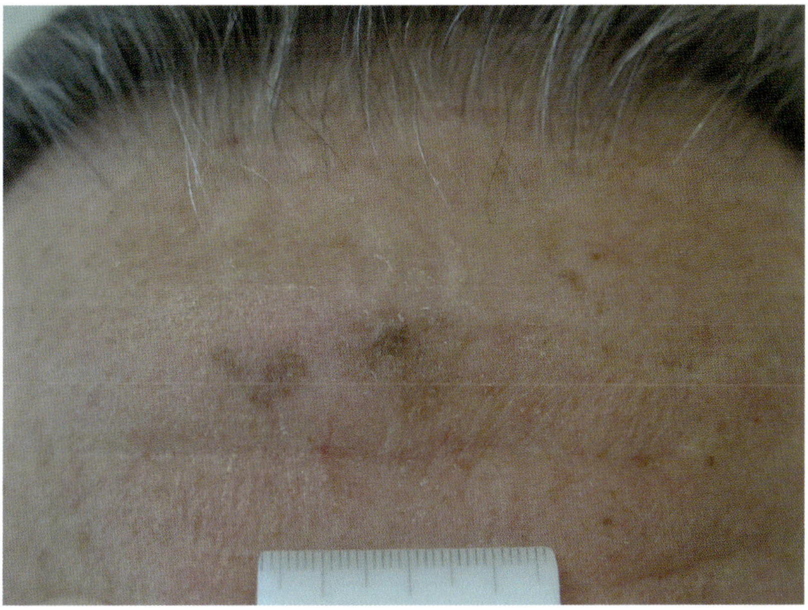

Fig. 3.1 The clinical examination of the forehead area revealed a poorly demarcated macule measuring several centimeters. The color was light-brown and focally darker in some areas. Slight desquamation was also present

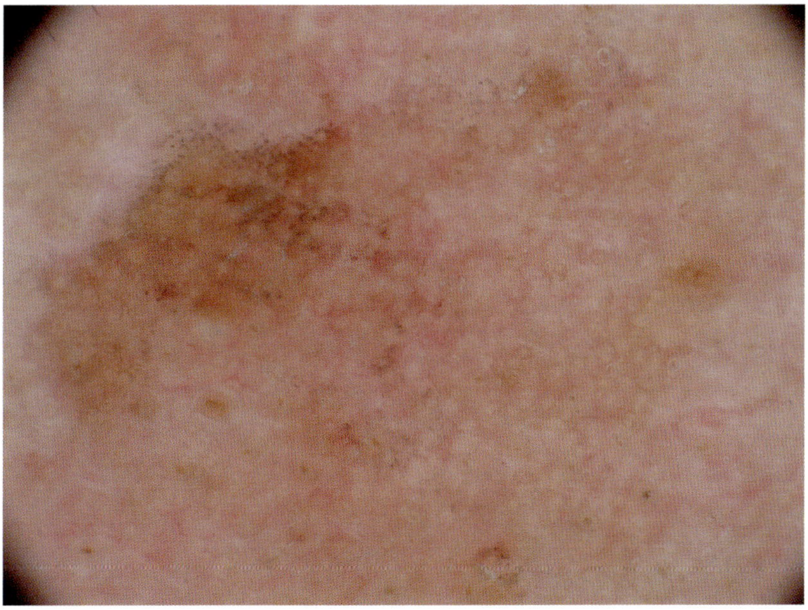

Fig. 3.2 Dermoscopy revealed an uneven distribution of colors and structures. Multiple gray-colored granules and dots were present mainly at the upper part of the lesion. Some dark-brown dots/globules were evident, too. At the left part, light-brown-colored circles and parallel lines are also seen

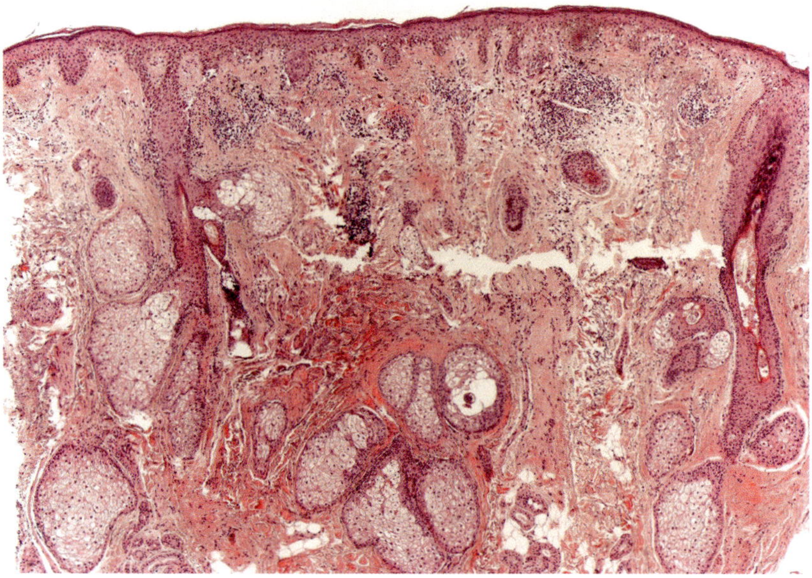

Fig. 3.3 Histologically, in the background of a severely sun-damaged skin, there was a monomorphic, quite atypical melanocytic proliferation, basally located and arranged in single cells

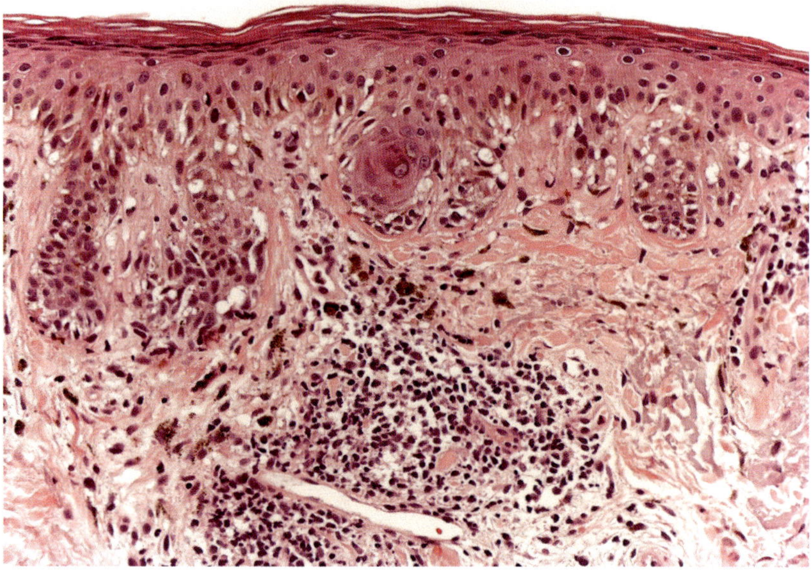

Fig. 3.4 Atypical melanocytes tended to be continuous and follow the adnexal epithelium

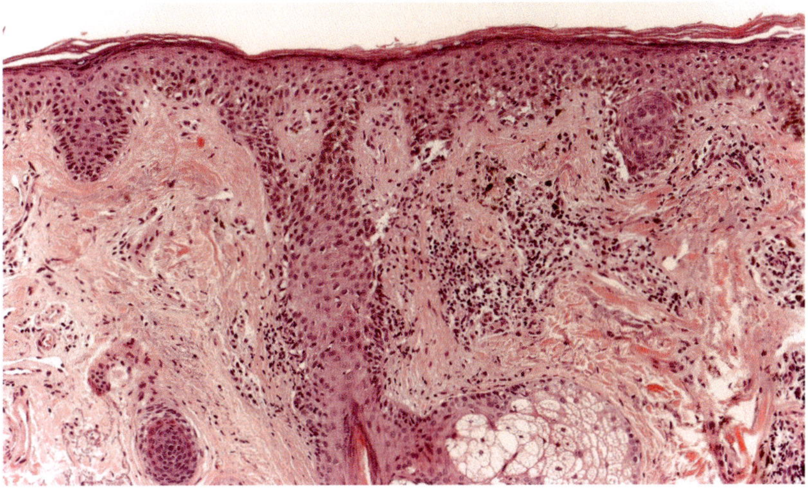

Fig. 3.5 Many melanophages and some lymphocytes were present in the elastotic dermis

Case 2

Patient: Male, 65 years old

Anatomic site of the lesion: Neck

History of the lesion: Since few years

Management: The dermoscopic pattern was quite unspecific. The differential diagnosis of flat pigmented facial lesions includes lentigo melanoma (LM), pigmented actinic keratosis (PAK), and solar lentigo/seborrheic keratosis (SL/SK). No feature suggestive of PAK or SL/SK could be seen. In addition, the presence of heavy pigmentation and the impeded visualization of the follicular openings were rather suggestive of LM. The lesion was excised to rule out melanoma.

Histopathologic diagnosis: Very early in situ melanoma

Key message: Epidemiological evidence suggests that flat and heavily pigmented nevi on the face of elderly individuals are extremely rare. The only nevus types that may be frequently seen at that age on the facial skin are dermal and blue nevi. Therefore, the differential diagnosis of a flat pigmented facial lesion includes LM, PAK, and SL/SK. The clinico-dermoscopic morphologic aspect of this lesion was not so compatible with PAK or SL/SK. Therefore, there are two possibilities: either this was a rare case of a flat nevus persisting in the seventh decade of life or this was a very early melanoma not having developed enough histopathologic criteria. In this context, the most reasonable choice is to accept the worst possibility and excise to safe margins.

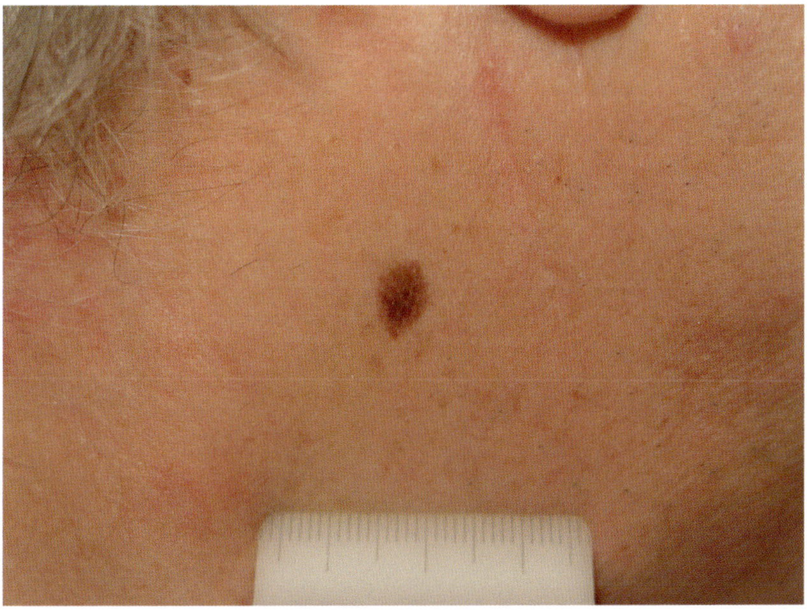

Fig. 3.6 A flat pigmented lesion on the neck with a maximum diameter of 8 mm. The lesion is relatively uniform in shape and color, with a slightly evident eccentric hyperpigmentation

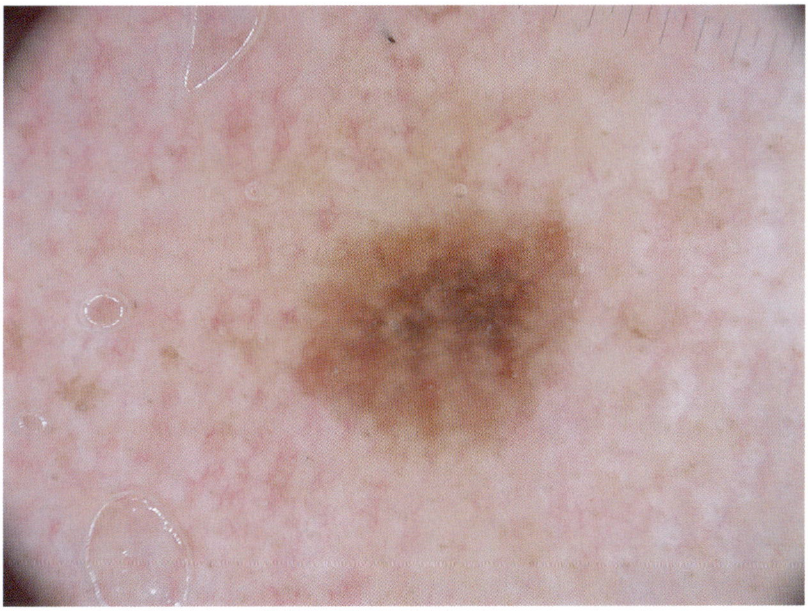

Fig. 3.7 Dermoscopy revealed mainly brown color, focally darker-brown, and maybe a few grayish globules. The visualization of the openings of the hair follicles is focally impeded by the presence of pigmentation

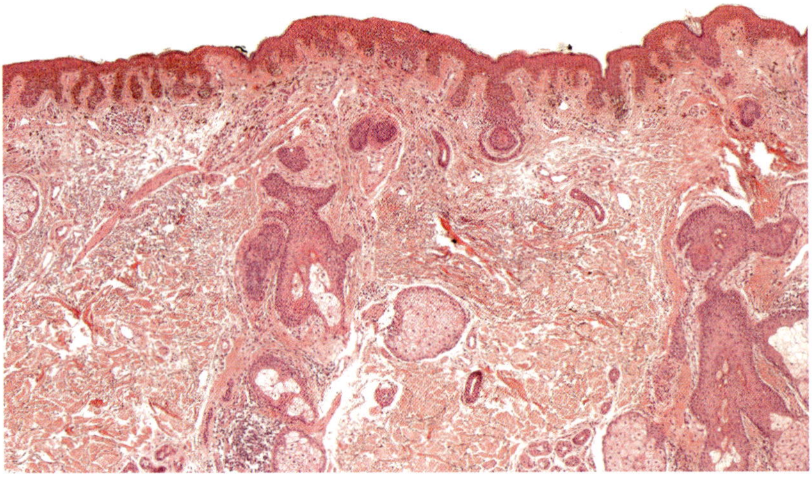

Fig. 3.8 On histology, the lesion was a very thin melanocytic proliferation, hypopigmented

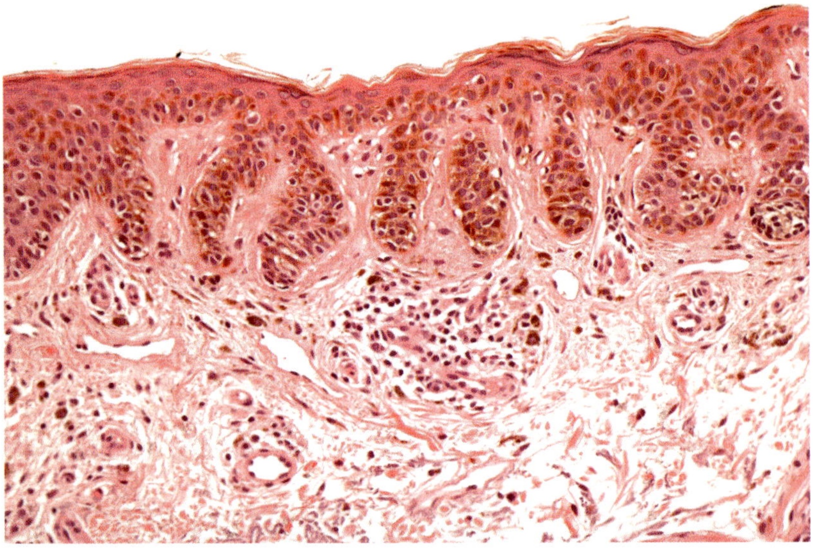

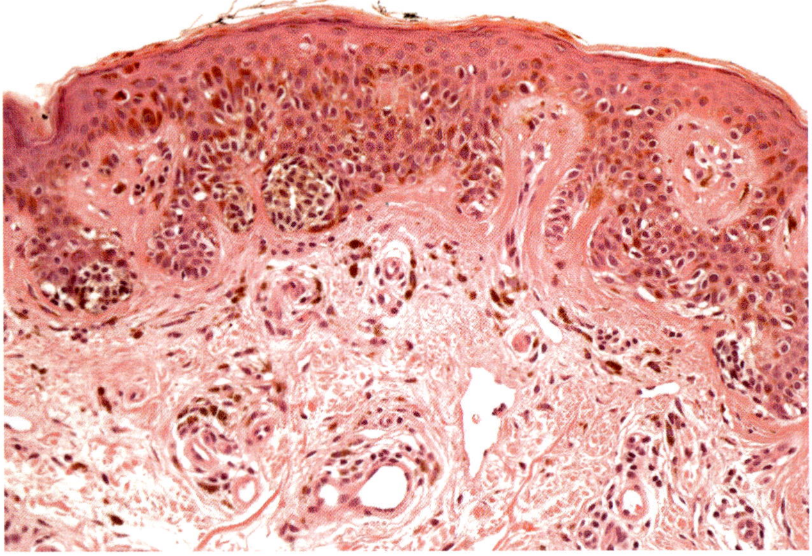

Figs. 3.9 and 3.10 Melanocytes were monomorphous, small, and unevenly distributed along the basal layer of the epidermis

Case 3

Patient: Female, 70 years old

Anatomic site of the lesion: Left cheek

History of the lesion: Since several years

Management: The clinical and dermoscopic image were not highly specific. The differential diagnosis included lentigo maligna, seborrheic keratosis, and pigmented actinic keratosis. However, the presence of few globules raised the suspicion of lentigo maligna. Since the lesion was not so large, a full elliptical excision was performed.

Histopathologic diagnosis: In situ melanoma (lentigo maligna) in sun-damaged skin

Key message: The present case has some dermoscopic features that can raise the suspicion of lentigo maligna. First of all, the presence of brown globule is highly suggestive for melanocytic proliferation. Second, the presence of polygons has been related to melanoma diagnosis. Third, there were no signs for other non-melanocytic proliferations. In fact, jelly borders, fingerprinting, rosettes, or comedo-like openings suggestive for solar lentigo, actinic keratosis, and seborrheic keratosis were not observed. Thus, the diagnostic approach of pigmented facial macules is to biopsy any given lesion revealing at least one dermoscopic features for melanocytic proliferation.

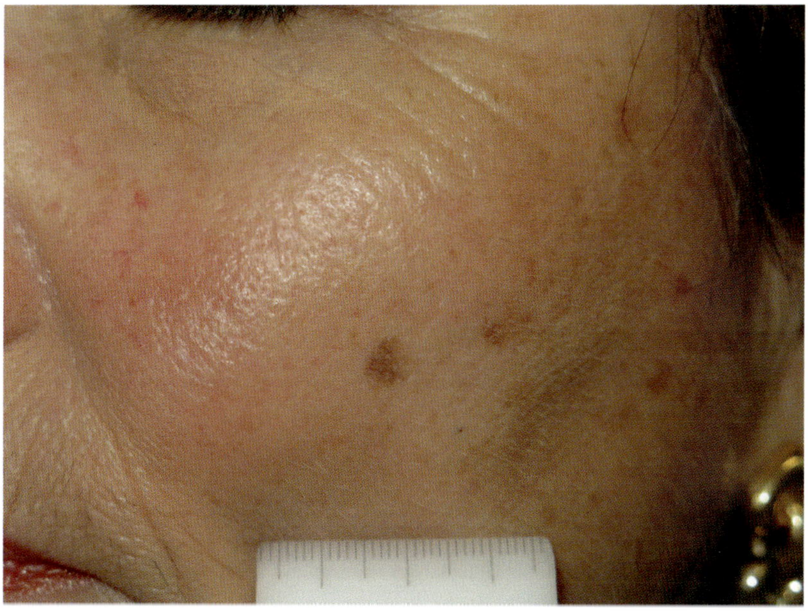

Fig. 3.11 The clinical examination of the left cheek revealed a poorly demarcated light-brown macule measuring 1 cm; the borders were irregular and the surface was not scaly

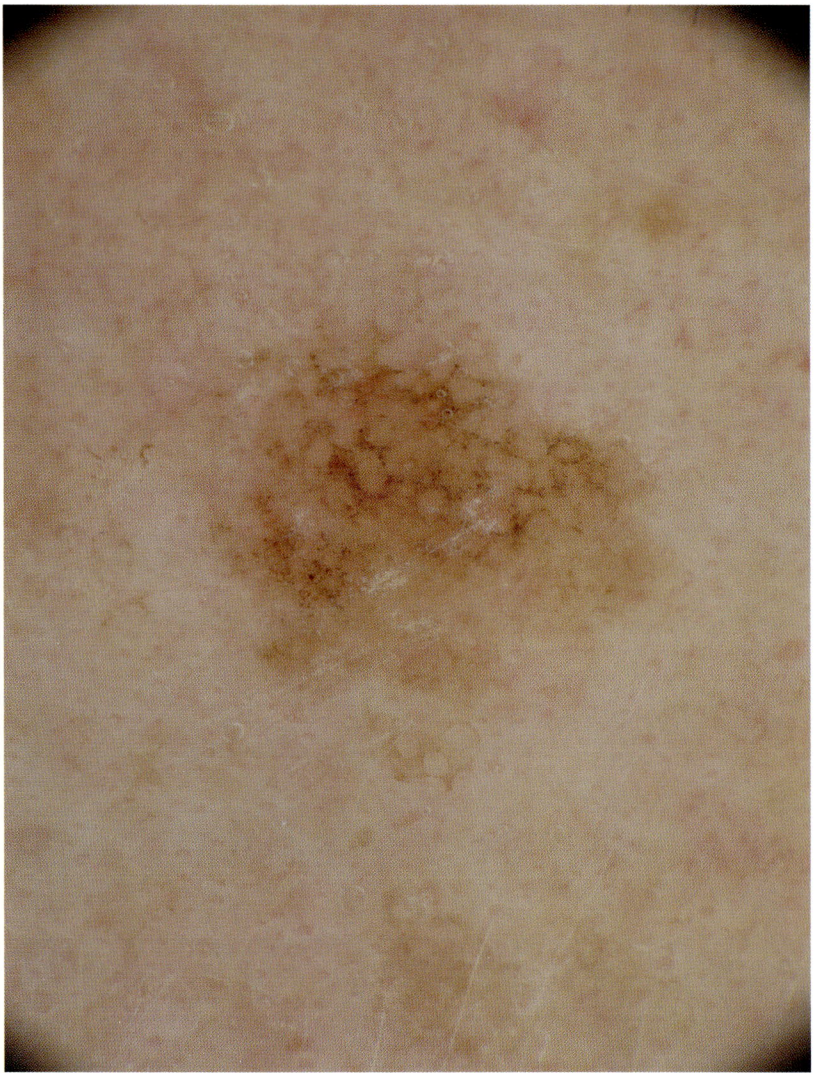

Fig. 3.12 Dermoscopy revealed the presence of multiple gray-colored granules and dots arranged to form polygons. At the intersection of those polygons, some dark-brown globules were seen

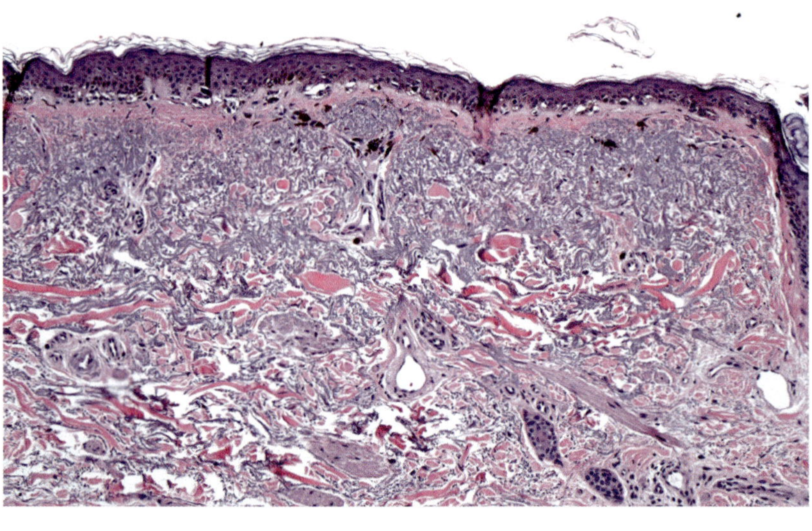

Fig. 3.13 In the background of severe actinic damage, an atypical melanocytic proliferation is growing at the tip of the epidermal papillae

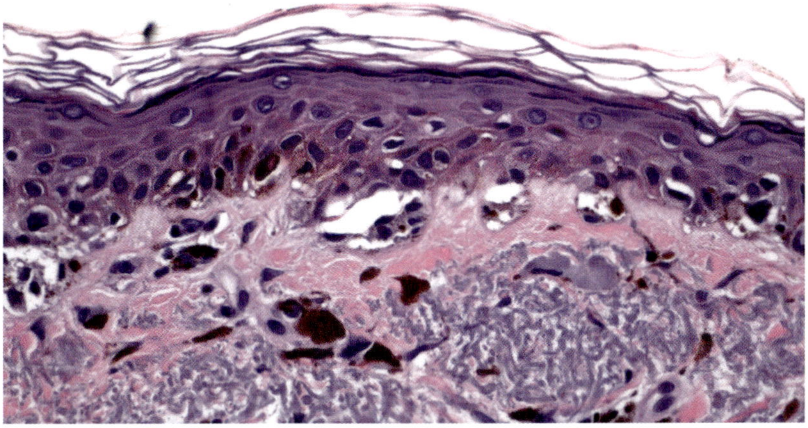

Fig. 3.14 Melanocytes are pleomorphic, sometimes with a clear perinuclear halo, always located in the basal epidermal layer. Melanophages in the superficial dermis contribute to the clinical pigmentation of the lesion

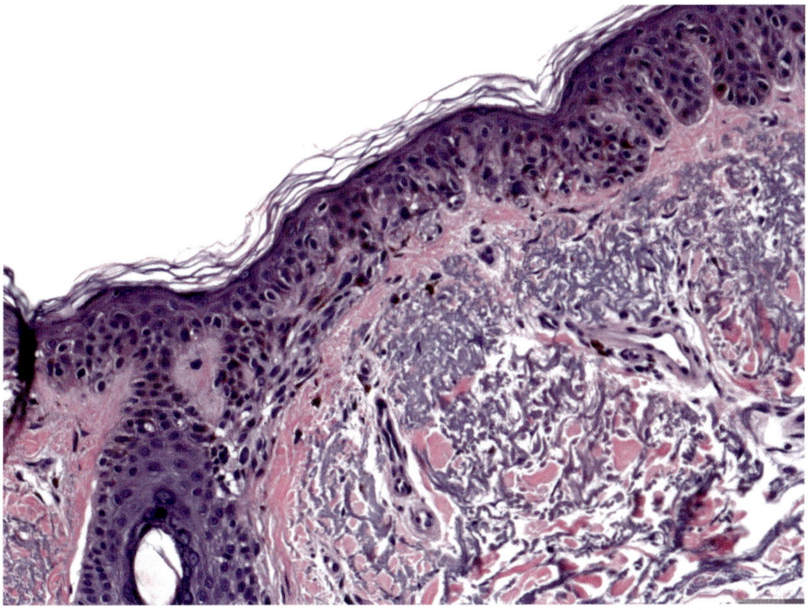

Fig. 3.15 Neoplastic melanocytes grow along the epidermal junction and involve also the follicular epithelium

Case 4

Patient: Male, 46 years old

Anatomic site of the lesion: Lower right eyelid

History of the lesion: Slowly growing persistent lesion

Management: Overall, all findings were suggestive for the diagnosis of melanoma.

Since the lesion was large, a punch biopsy was preferred to a full elliptical surgical excision. Neither the dimension of the lesion nor the clinical suspect was transmitted to the dermatopathologist.

First histopathologic diagnosis: Histopathologic examination made on punch biopsy, without knowing any clinical information, was of compound nevus with inflammation.

Second histopathologic diagnosis after clinicopathologic consultation: Superficial spreading melanoma, Breslow thickness 0.3 mm

Key message: This case highlights well the need of a close collaboration between clinicians dealing with the patient and pathologists looking through a light microscope. Although histopathology is regarded as the golden standard, several diagnostic difficulties could be encountered at different points. In particular, a small biopsy may not reveal the diagnostic clues especially for the possible occurrence of a discontinuous melanocytic proliferation (skip areas) and the fact that could be very difficult the distinction of atypical melanocytes of LM from the pleomorphic, atypical melanocytes in sun-damaged skin. On the other side, clinicians should be aware that a histologic diagnosis of "junctional nevus" in adults cannot be accepted passively since it is well-known that nevi of the head/neck of adults are nodular and commonly reveal histopathologic features of a benign intradermal nevus. Furthermore, the large size of the lesion should raise the index of suspicion since junctional nevi are usually smaller in size and thus the large size of the lesion of this case is suggestive of growth. As a general rule, a growing melanocytic lesion is rather unusual in adults and should always raise the suspicion of melanoma, regardless of the body location.

The take-home message of this case is that it is mandatory for clinicians to critically review any diagnosis of a junctional or atypical lentiginous nevus on the head/neck region. Furthermore, they have to provide all clinical information of sex, age, body location, and possibly clinical and dermoscopic pictures to the pathologist who receives the specimen.

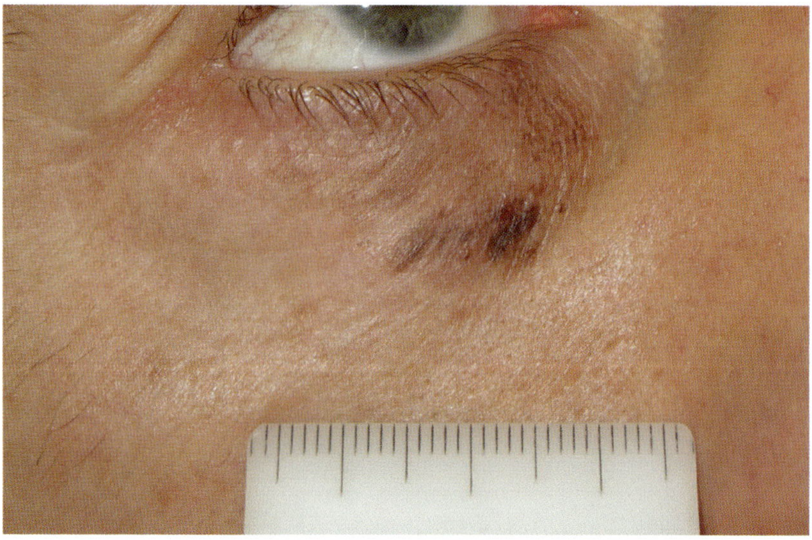

Fig. 3.16 Clinically the lesion was a light-brown flat patch, measuring 2.5 cm in its large diameter and with irregular border

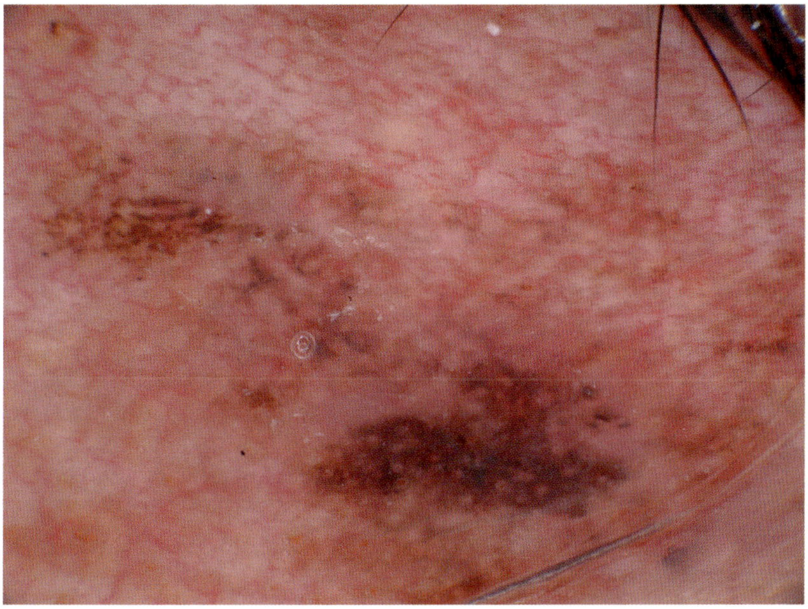

Fig. 3.17 Dermoscopically, an atypical network with variegated colors ranging from light- to dark-brown and with blue-gray color was detected. Furthermore, pigmented globules were seen at the periphery of the lesion

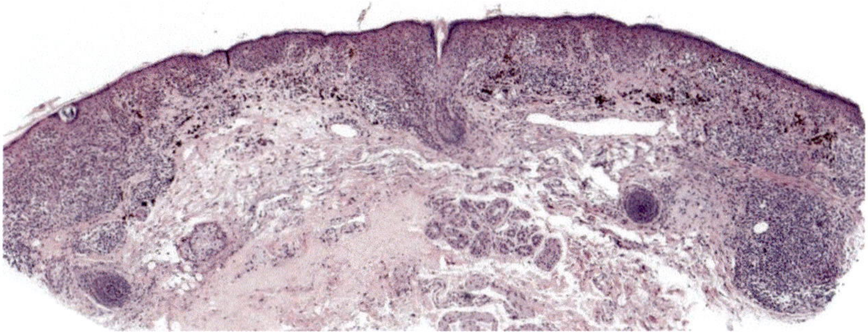

Fig. 3.18 At scanning view, the punch biopsy is completely occupied by a melanocytic proliferation that, at the borders, seems to involve the dermis. The actinic damage is diffuse

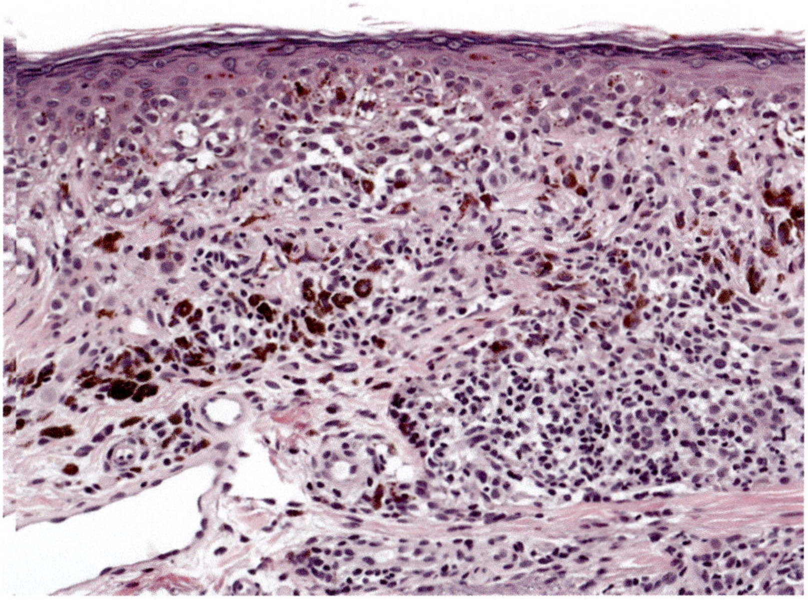

Fig. 3.19 Epidermal melanocytes are mostly located at the junction and have a rather monomorphous appearance. Many melanophages and a brisk lymphocytic infiltrate can be seen in the superficial dermis

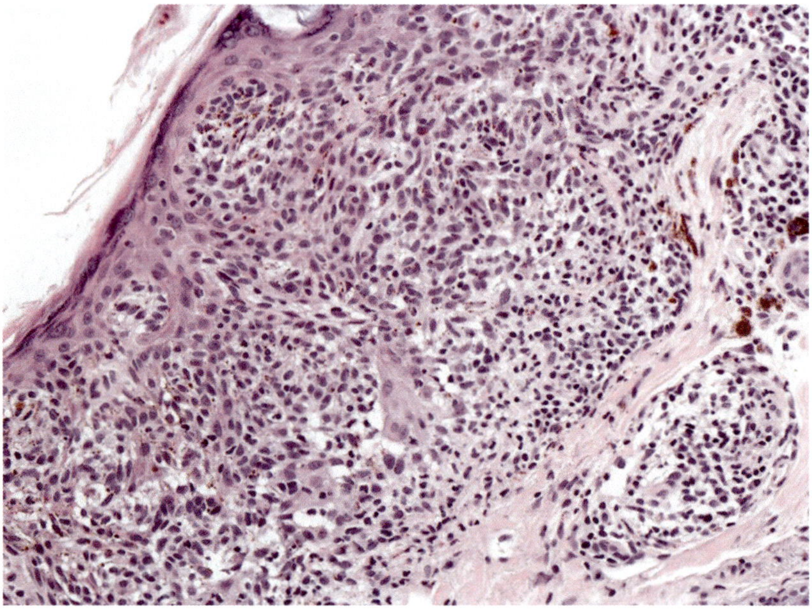

Fig. 3.20 Junctional melanocytes are vaguely spindle and benign-looking

Case 5

Patient: Male, 66 years old

Anatomic site of the lesion: Nose

History of the lesion: Unknown

Management: The unusual body site location and the presence of gray color raised the suspicion for melanoma diagnosis, and thus a biopsy was performed.

Histopathologic diagnosis: In situ melanoma on sun-exposed skin

Key message: Lentigo maligna is commonly a large flat lesion arising in elderly. In this case, the lesion was small in size, but it displayed all dermoscopic features that are suggestive for melanoma diagnosis such as the presence of gray circles. Thus, dermoscopy is useful not only for the diagnosis of "classic" large lentigo maligna but also for incipient small tumors.

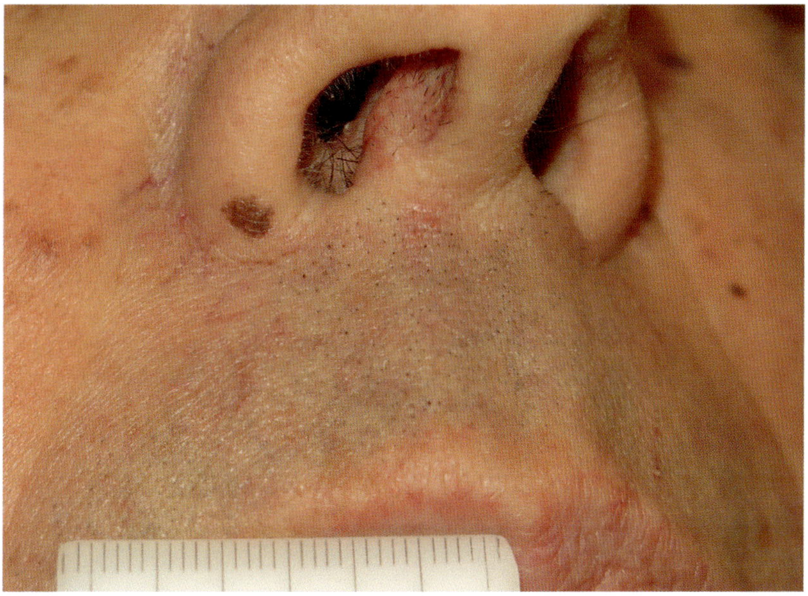

Fig. 3.21 The clinical examination of the lower area of the right pinna revealed a small (0.4 × 0.4 cm) flat pigmented lesion

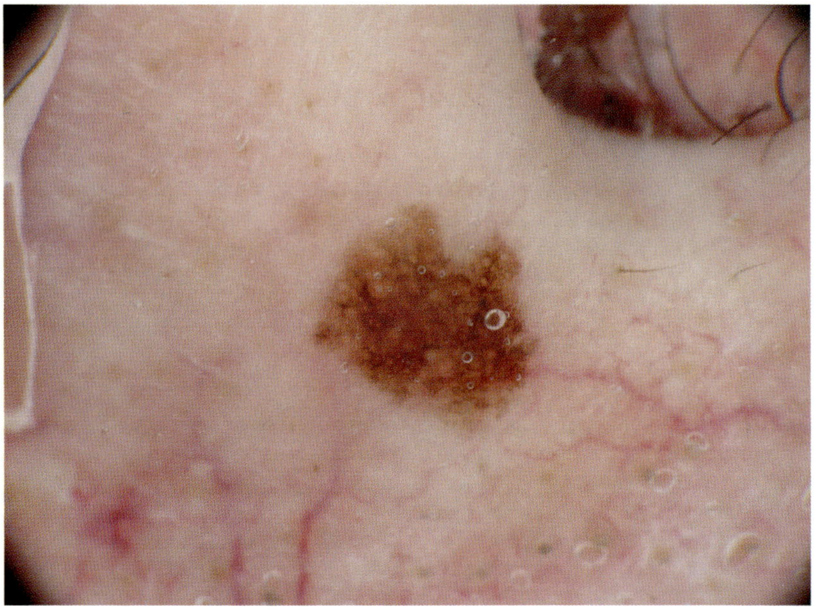

Fig. 3.22 Dermoscopy revealed an uneven distribution of pigment around the hair follicles (gray circles) and fading off borders

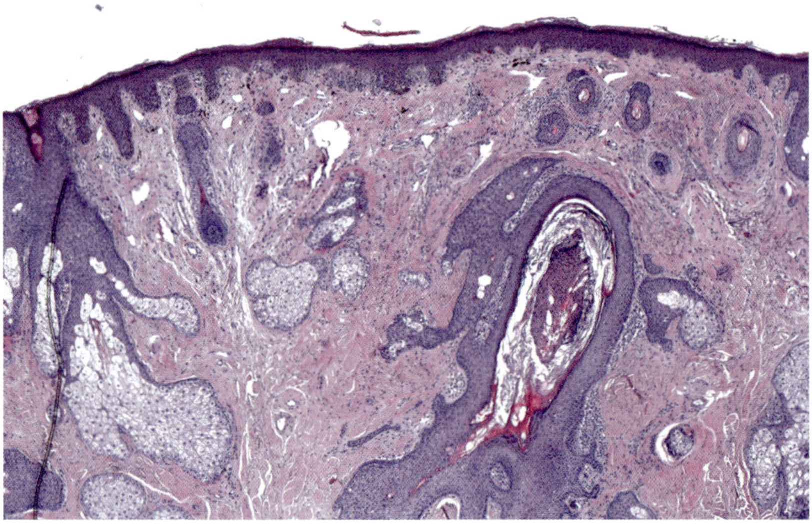

Fig. 3.23 At low-power view, a subtle melanocytic proliferation is present at the dermoepidermal junction

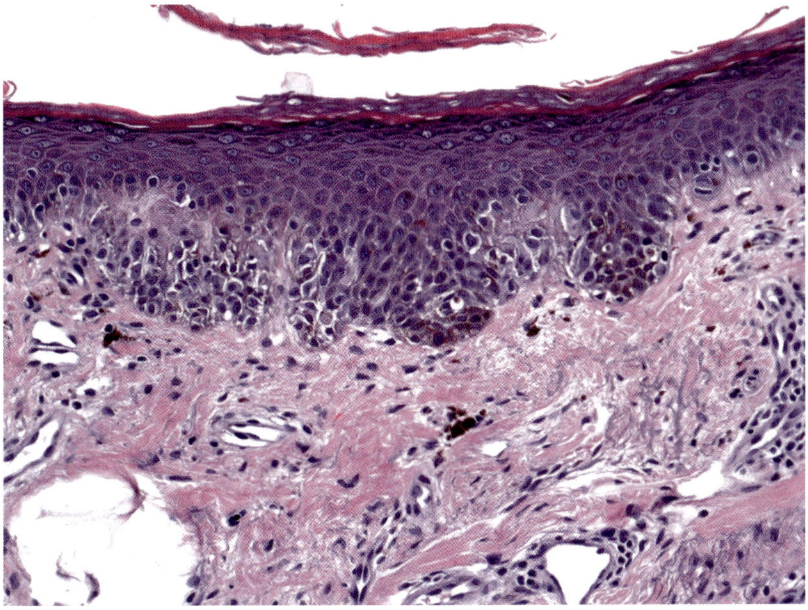

Fig. 3.24 Melanocytes are atypical and slightly epithelioid but always basally located

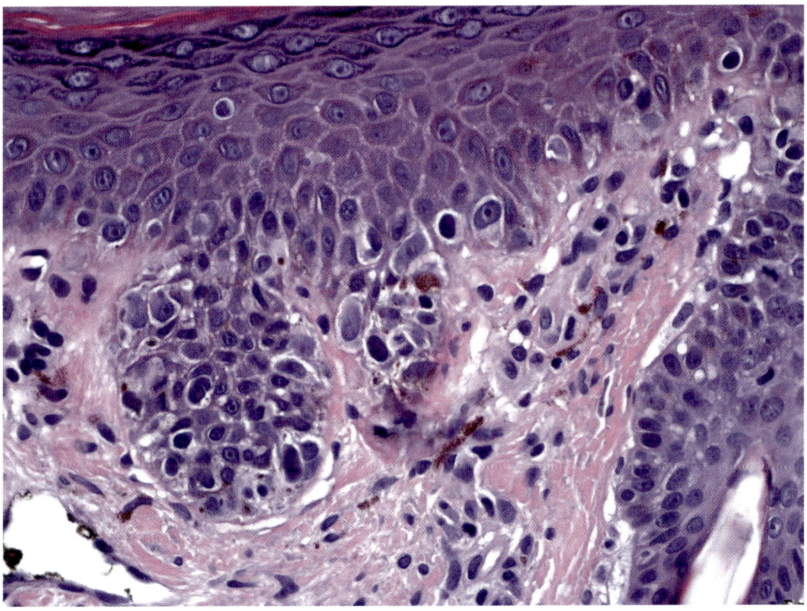

Fig. 3.25 Hyperchromasia, nucleomegalia, and evident nucleoli are malignant cytological features

Suggested Reading

1. Lallas A, Argenziano G, Moscarella E, et al. Diagnosis and management of facial pigmented macules. Clin Dermatol. 2014;32:94–100.
2. Lallas A, Tschandl P, Kyrgidis A, et al. Dermoscopic clues to differentiate facial lentigo maligna from pigmented actinic keratosis. Br J Dermatol. 2016;174:1079–85.
3. Zalaudek I, Lallas A, Longo C, et al. Problematic lesions in the elderly. Dermatologic Clinics 2013; 31:549–64–vii–viii.
4. Tiodorovic-Zivkovic D, Argenziano G, Lallas A, et al. Age, gender, and topography influence the clinical and dermoscopic appearance of lentigo maligna. J Am Acad Dermatol 2015;72:801–8.
5. Peris K, Maiorino A, Di Stefani A, Longo C, Piana S, Argenziano G. Brown globules in lentigo maligna (LM): a useful dermoscopic clue. J Am Acad Dermatol. 2016;75(2):429–30.
6. Zalaudek I, Cota C, Ferrara G, Moscarella E, Guitera P, Longo C, Piana S, Argenziano G. Flat pigmented macules on sun-damaged skin of the head/neck: junctional nevus, atypical lentiginous nevus, or melanoma in situ? Clin Dermatol. 2014;32(1):88–93.
7. Ferrara G, Argenziano G, Giorgio CM, Zalaudek I, Kittler H. Dermoscopic-pathologic correlation: apropos of six equivocal cases. Semin Cutan Med Surg. 2009;28(3):157–64.
8. Pralong P, Bathelier E, Dalle S, Poulalhon N, Debarbieux S, Thomas L. Dermoscopy of lentigo maligna melanoma: report of 125 cases. Br J Dermatol. 2012;167(2):280–7.
9. Reed JA, Shea CR. Lentigo maligna: melanoma in situ on chronically sun-damaged skin. Arch Pathol Lab Med. 2011;135(7):838–41.

Recurrent Nevi and Nevi with Sclerosing Features and Inflammation

4

Recurrent nevi are benign melanocytic nevi that regrow after incomplete surgical removal or traumatic injury. Recurrent nevi are biologically benign without a documented risk of malignant progression, but they pose diagnostic problems because their morphological features resemble those of melanoma. They are also commonly referred to as "pseudomelanomas." The management of repigmentation within a scar after incomplete biopsy or other destructive procedures (i.e., laser and cryotherapy) depends on whether a prior histopathologic diagnosis confirming a benign tumor was obtained or not. In cases of regrowth without histopathologic diagnosis, immediate excision and histopathologic examination are mandatory to rule out melanoma unintentionally treated as benign tumor. When a prior histopathologic diagnosis reports a benign tumor, often no further interventions are performed. However, careful histologic reevaluation of the slides of the previous excision is strongly encouraged, in order to avoid a misclassification which potentially delays diagnosis and adequate treatment of recurrent melanoma. The differential diagnosis of newly developing pigmentation within a scar of a previous excision of a melanocytic lesion includes recurrent nevus, recurrent melanoma, and melanotic reactive pigmentation. As a general rule, a relatively rapid development of the repigmentation favors the diagnosis of a benign recurrence. This is confined within the scar, and, dermoscopically, a regularly pigmented network and thin parallel streaks are associated with reactive pigmentations, whereas an irregular prominent network, globules, streaks, and heterogeneous pigmentation are usually associated with a melanocytic proliferation. On the other side, melanomas tend to recur months if not years after excision beyond the scar. Dermoscopy may not reveal specific melanoma features, displaying only a heterogeneous, structureless pigmentation. Therefore, in the case of a recurrent pigmentation involving the normal skin around the scar, a biopsy specimen should be obtained to exclude a possible melanoma recurrence. Histological features, even in case of recurrent nevi, can be very worrisome as recurrent nevi generally present a rich junctional component, sometimes frankly atypical and with epidermotropism, arising on a sclerotic dermis.

Sclerosing nevus with pseudomelanomatous features (or nevus with regression-like fibrosis) is a histopathologic entity that can closely simulate regressing melanoma. These nevi mainly occur on the back of young individuals. They are thought to be the result of minor, unnoticed trauma on preexisting nevi and are histopathologically characterized by an evident

C. Longo et al., *Atlas of Diagnostically Challenging Melanocytic Neoplasms*,
https://doi.org/10.1007/978-3-319-48653-6_4

junctional proliferation associated with some pagetoid spreading, areas of dermal fibrosis/sclerosis containing architecturally altered melanocytic nests, and residual nevus tissue (often with congenital-like features) around and deep into the cicatricial tissue. Sometimes the histological differential diagnosis with a melanoma can be very challenging; regressing melanoma usually has clear cytological atypia, both in the junctional and in the dermal component, and can show expansive dermal nodules of neoplastic melanocytes with or without tumoral melanosis. All these features lack in sclerosing nevi.

Clinically and dermoscopically, they show atypical overall features, and regression is extensive and polychromatic. However, the clinical information of a young individual with multiple similar looking lesions on the back can be an important information favoring a benign diagnosis. In the case of solitary lesions, excision is mandatory as histopathology remains the gold standard for diagnosis. In this chapter we try to elucidate some of the challenges of these peculiar situations.

Case 1

Patient: Female, 50 years old

Anatomic site of the lesion: Upper back

History of the lesion: The patient had a few moles on the trunk and the extremities. She referred a previous excision of a dermal nevus on the same location. She noticed the appearance of this irregular pigmentation a few months after the biopsy.

Management: The lesion was excised to rule out recurrent melanoma.

Histopathologic diagnosis: Recurrent nevus

Key message: Both melanomas and nevi can recur after incomplete excision. The differential diagnosis between recurrent nevus and recurrent melanoma can be challenging from a clinical and dermoscopic point of view. As a general rule, recurrent nevus does not grow outside the scar of the previous excision and tends to regrow from the center to the periphery of the scar.

Even if histologically the vast majority of recurrent nevi are readily identifiable, a diagnosis of recurrent nevus should be done with caution and after the revision of the previous biopsy.

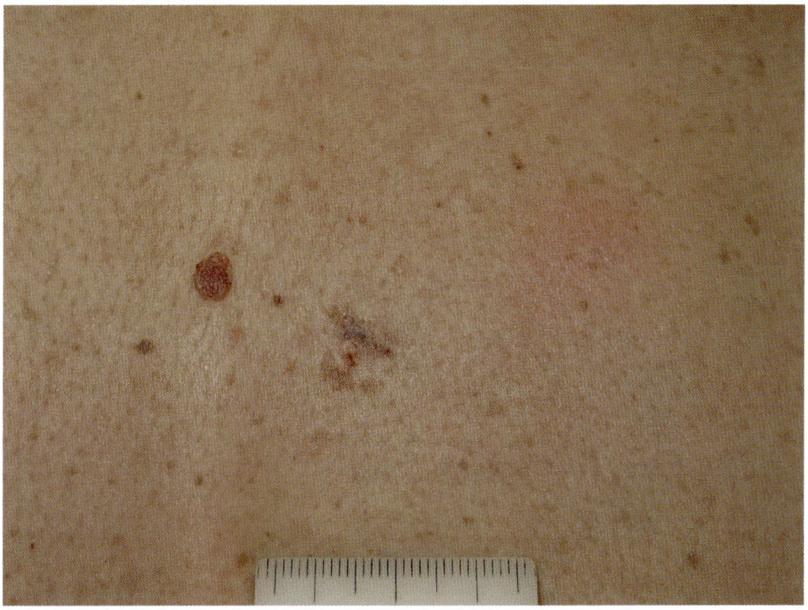

Fig. 4.1 Close-up clinical examination revealed an irregularly pigmented macule, 1 cm diameter, with uneven pigmentation and irregular borders

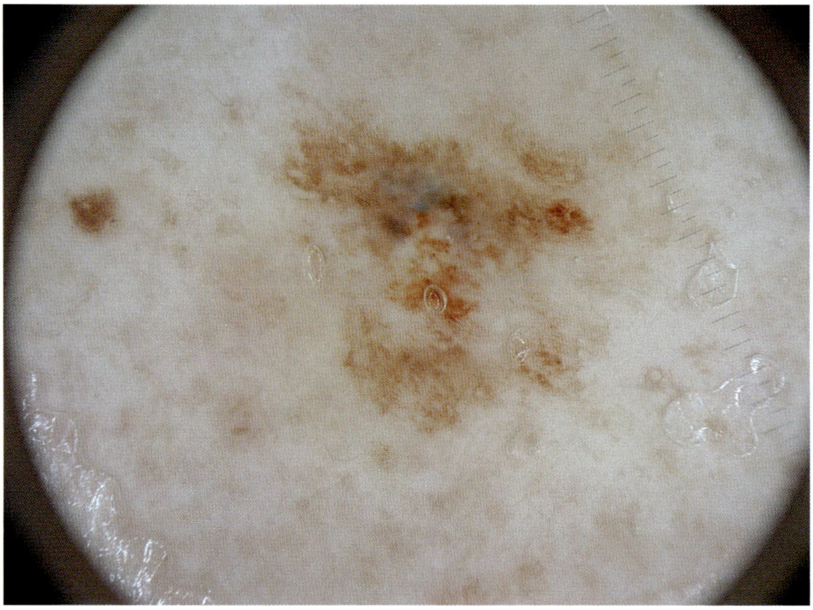

Fig. 4.2 Dermoscopically, the lesion was characterized by atypical pigment network at the periphery and regression structures in the center of the lesion, with white scar-like areas and bluish globules

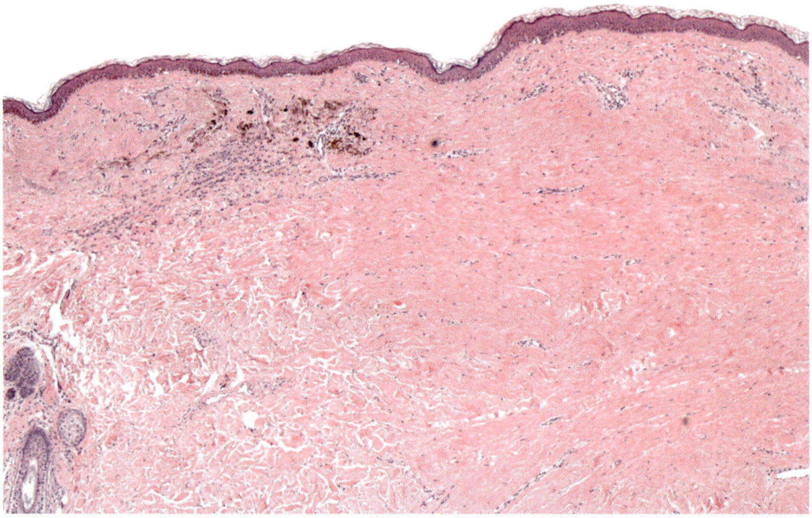

Fig. 4.3 At low-power view, the dermis showed a dense scarring (on the *right*) and some residual melanocytes (on the *left*)

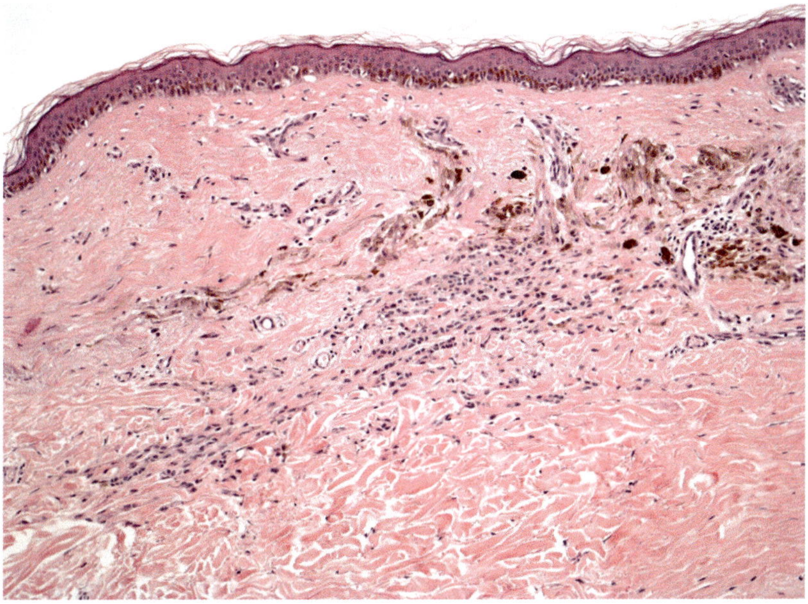

Fig. 4.4 The dermal melanocytes were cytologically bland; some melanophages were present

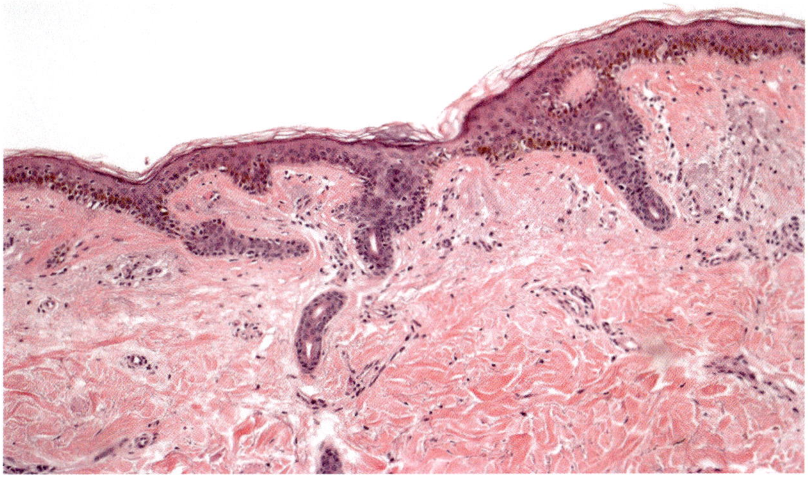

Fig. 4.5 The junctional melanocytic hyperplasia with partial effacement of the retiform epidermis is a well-defined phenomenon in recurrent nevi

Case 2

Patient: Male, 48 years old

Anatomic site of the lesion: Lower back

History of the lesion: Unknown

Management: The presence of dotted vessels in a flat nonpigmented lesion is suggestive of a melanocytic lesion. Although nonpigmented nevi do exist, this was the only lesion deviating the "signature" pattern of nevi of the patient, which were all pigmented and reticular on dermoscopy. Therefore, the lesion was excised to rule out melanoma.

Histopathologic diagnosis: Compound nevus with dermal inflammation

Key message: Halo phenomenon is well-known to occur in melanocytic tumors. In children, the development of usually multiple halo or Sutton nevi is quite a frequent finding. In contrast, solitary lesions with whitish halos in adults should be excised, since the same phenomenon has been described to occur in melanomas, resulting in a very similar dermoscopic appearance. This lesion finally turned out as a nevus, but this could not be established on a clinico-dermoscopic basis alone.

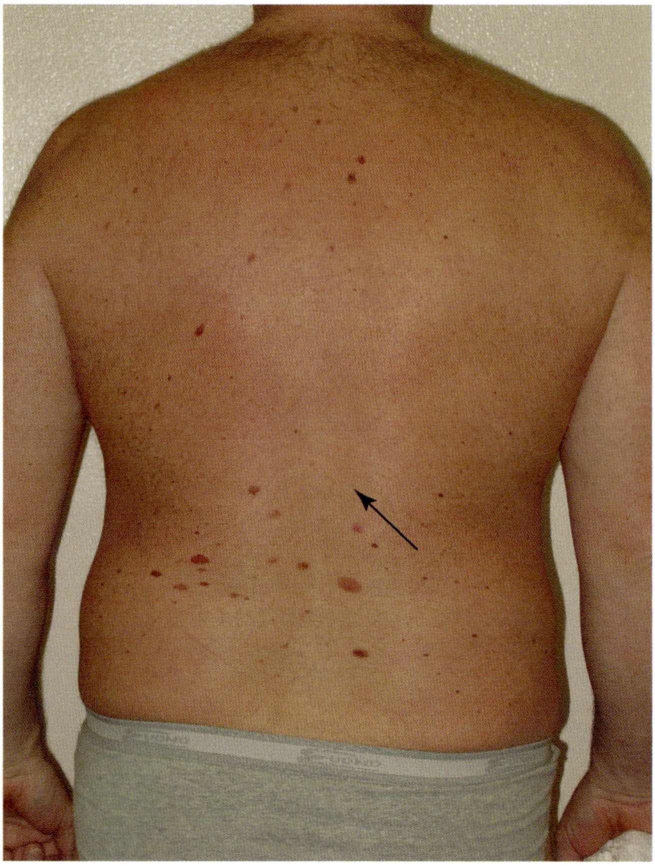

Fig. 4.6 The patient had several nevi on the back, some of them clinically "atypical"

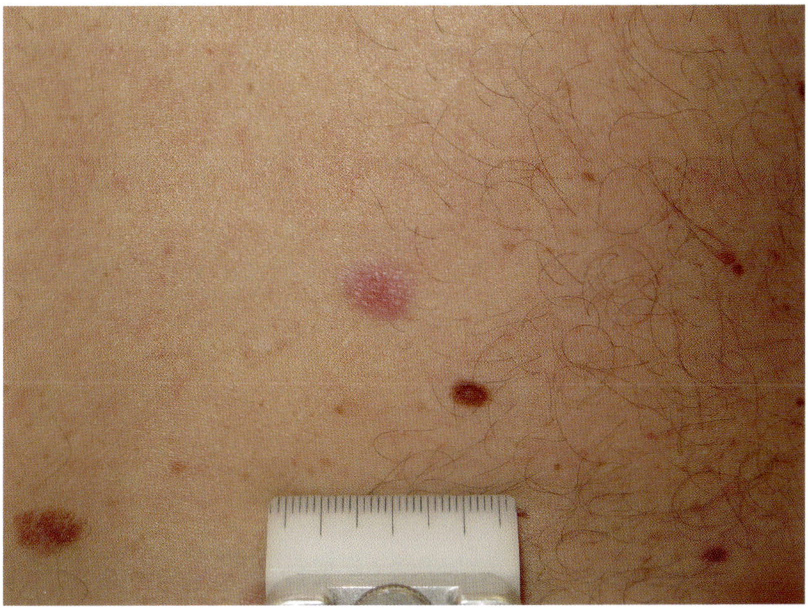

Fig. 4.7 Among several pigmented nevi, there was a nonpigmented (*pinkish*) flat lesion on the lower back of the patient

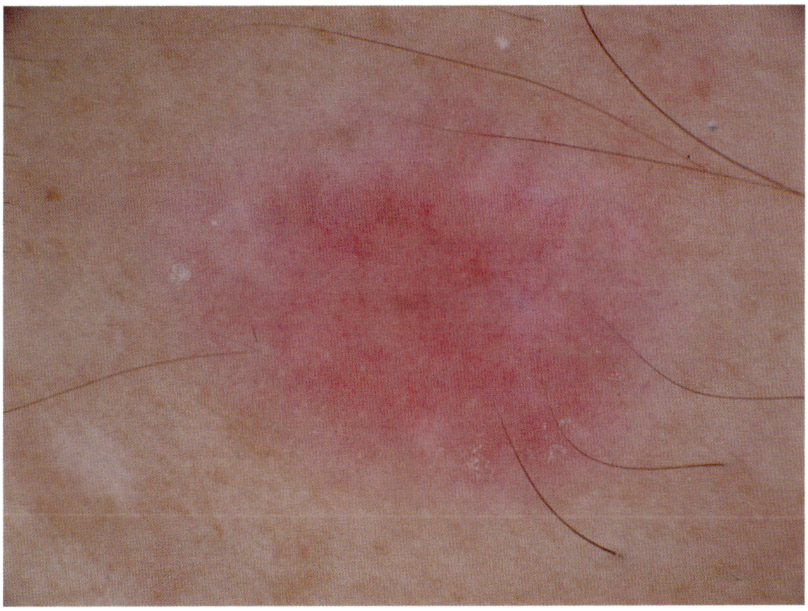

Fig. 4.8 Dermoscopy revealed multiple dotted and short linear curved (*comma-like*) vessels. A whitish hue was evident at the periphery of the lesion

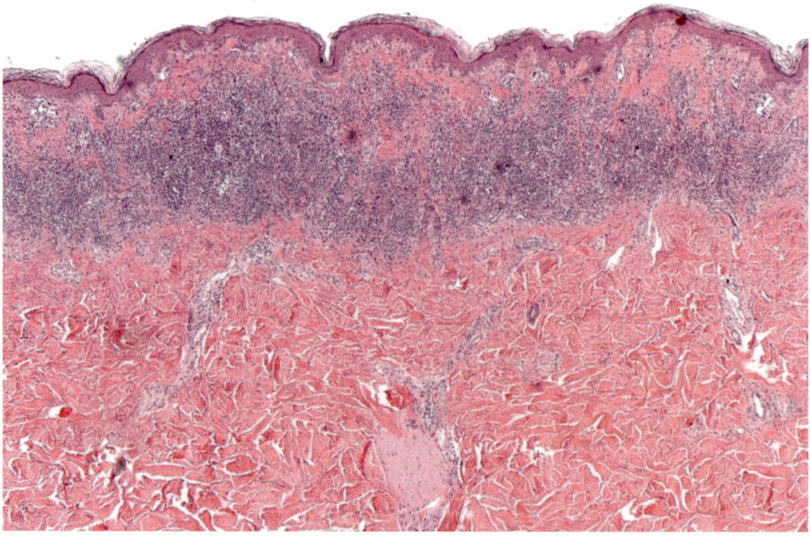

Fig. 4.9 At scanning view, a highly cellular lesion could be appreciated

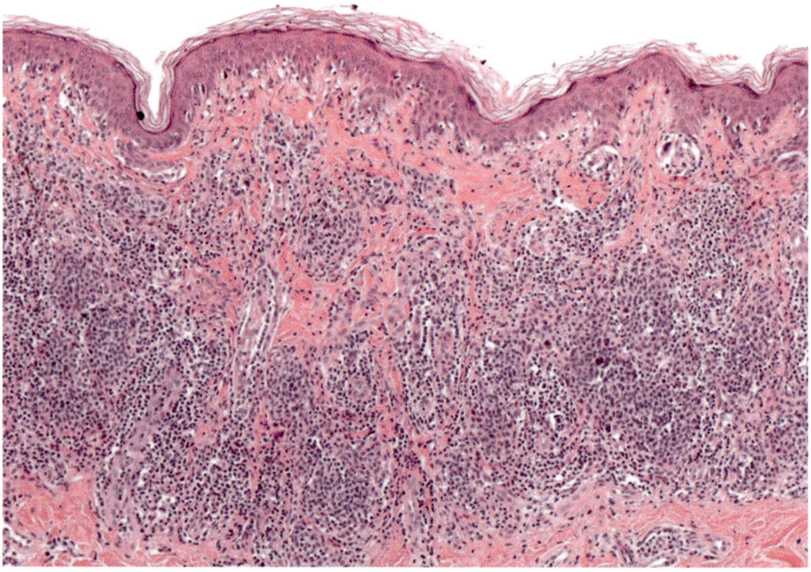

Fig. 4.10 At higher magnification, a residual focal junctional activity was present, but the lesion was almost completely dermal and was diffusely infiltrated by lymphocytes

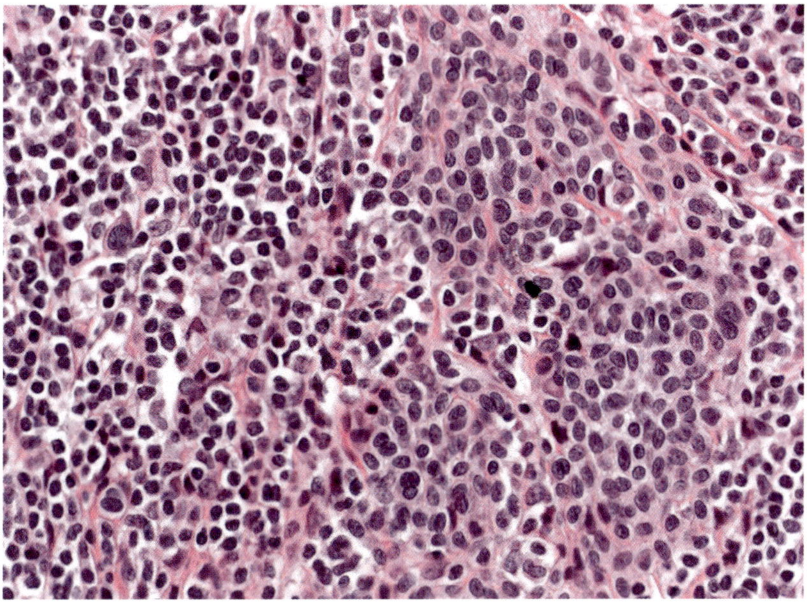

Fig. 4.11 The nevus cell population (right half of the figure) was briskly admixed with the inflammatory cells

Case 3

Patient: Male, 57 years old

Anatomic site of the lesion: Upper back

History of the lesion: The lesion appeared approximately 2 years ago and gradually enlarged.

Management: Based on the presence of extensive regression, the lesion was excised to rule out melanoma.

Histopathologic diagnosis: Compound nevus with dermal sclerosis

Key message: Sclerosing nevus with pseudomelanomatous features has been described as a peculiar subtype of Clark nevus, characterized by regression-like fibrosis. As a rule, the dermoscopic pattern of these nevi is very worrisome, since they display features typically associated with melanoma. Therefore, they are very often excised and recognized only histopathologically.

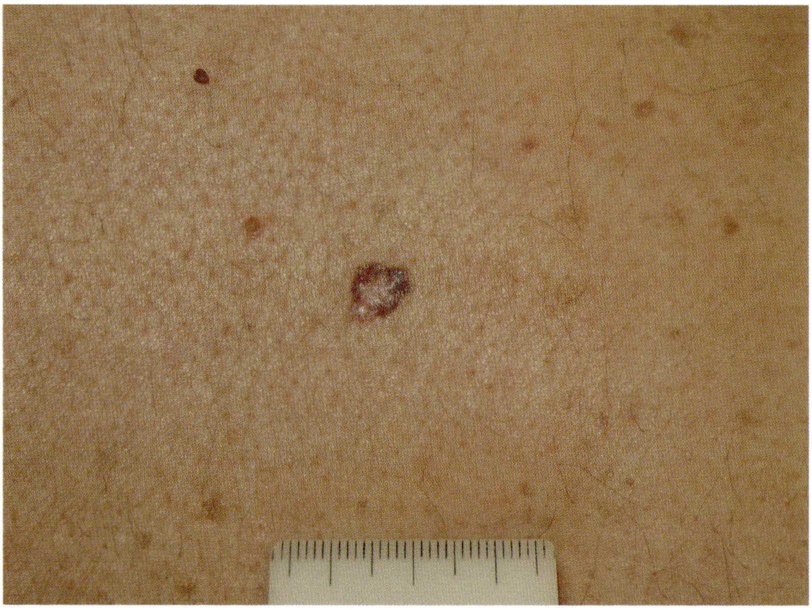

Fig. 4.12 A clinically atypical pigmented lesion, with a maximum diameter of almost 1 cm, characterized by multiple colors, evident regression in the center, and an irregular border

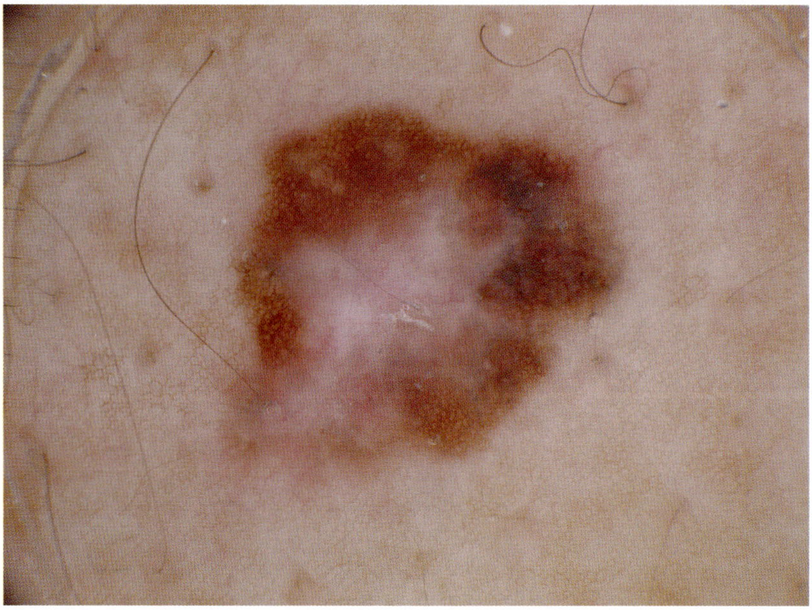

Fig. 4.13 Dermoscopy revealed extensive regression, typified by a white, scar-like area with telangiectatic vessels covering a large part of the lesion, as well as gray dots, globules, and blotches

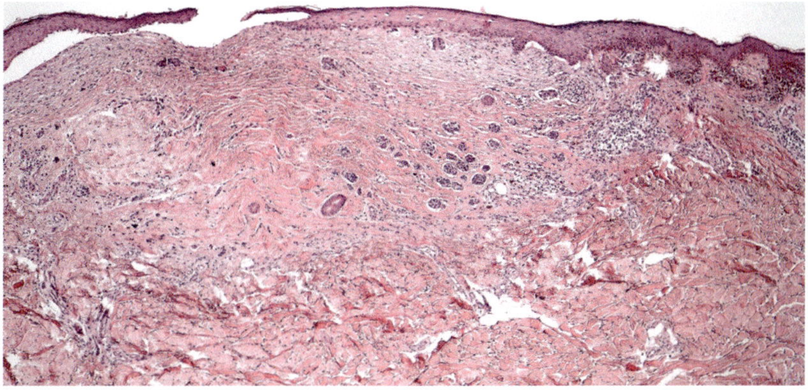

Fig. 4.14 The main histologic feature of the lesion was the dense, nodular, dermal fibrosis in which some small nests of mature melanocytes were involved

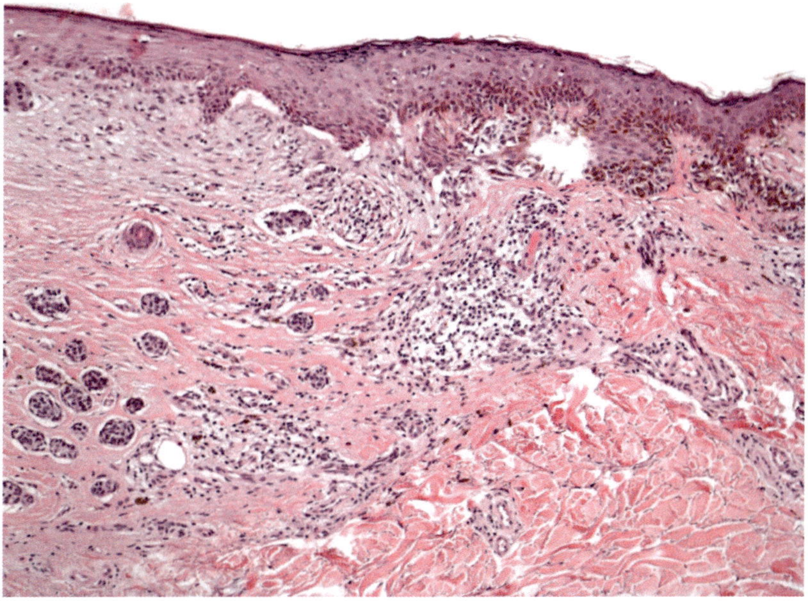

Fig. 4.15 At the border of the lesion, a junctional pigmented melanocytic proliferation was juxtaposed to the dermal component

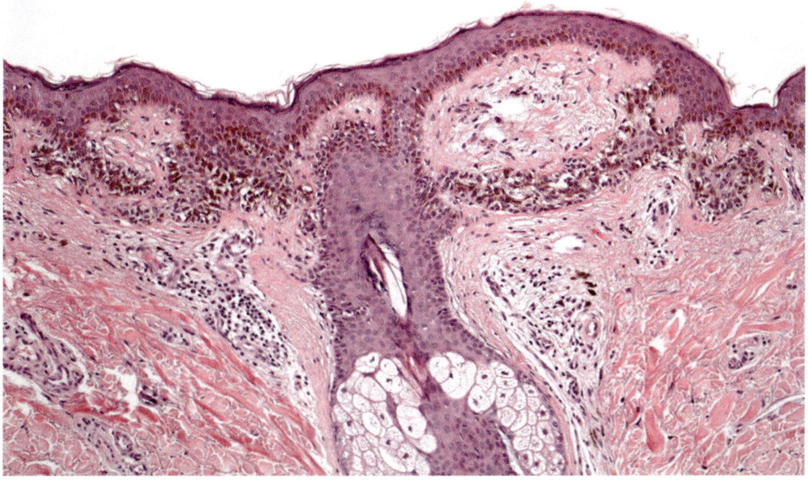

Fig. 4.16 The junctional proliferation had morphological features consistent with Clark nevus

Case 4

Patient: Male, 35 years old

Anatomic site of the lesion: Lower back

History of the lesion: Unknown

Management: Based on the marked clinical and dermoscopic asymmetry, the lesion was excised to rule out melanoma.

Histopathologic diagnosis: Combined nevus with dermal sclerosis

Key message: Dermoscopy improves melanoma recognition by revealing the morphologic asymmetry of the tumor before it becomes clinically evident. Lesions characterized by marked dermoscopic asymmetry and the presence of at least one of the so-called melanoma-specific criteria should be excised to rule out melanoma. However, the specificity of dermoscopy is not 100%, especially if clinicians lower the excision threshold in order not to miss early melanomas. Therefore, in the everyday clinical practice, some of the lesions dermoscopically judged as suspicious will turn out to be nevi histopathologically. In these cases, the histopathologic report should be interpreted within the context of the patient. In the present example, the age of the patient and the coexistence of several nevi with some atypia allowed to accept the diagnosis of a nevus, provided that the histopathologist was well-informed about the clinico-dermoscopic atypia.

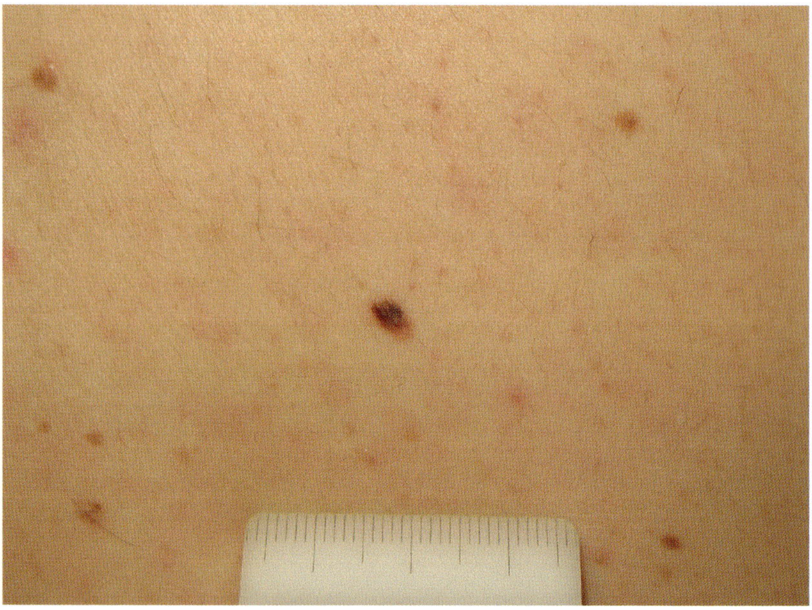

Fig. 4.17 Among several nevi on the trunk of this patient, there was a small but highly asymmetric mole on his lower back. The maximum diameter was 5 mm, but the lesion was characterized by a marked asymmetry of color, with a *dark-brown*-to-*black upper part* and a *light-brown*-to-*red lower part*

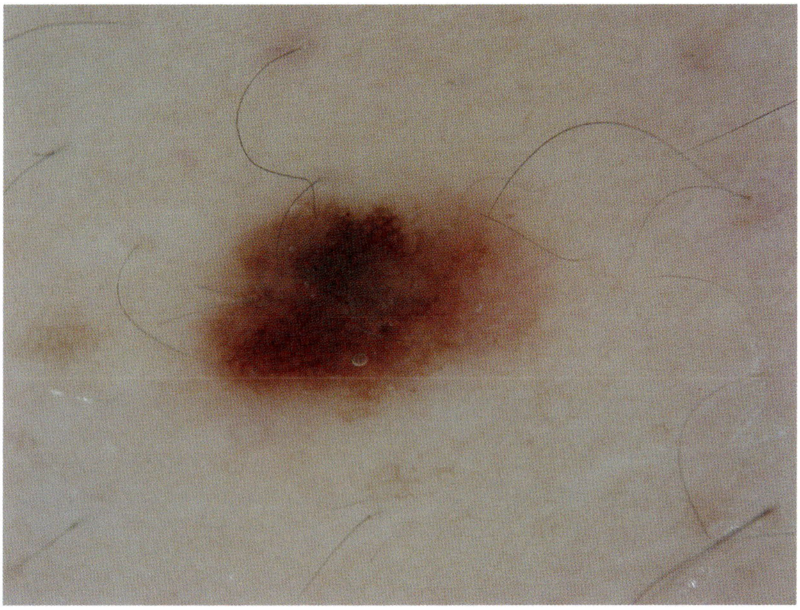

Fig. 4.18 The asymmetry of the lesion was further highlighted by dermoscopy, which revealed a significant heterogeneity of pigment distribution. The *dark-brown* part consisted of pigment network with thicker than usual lines and abrupt ending at the periphery (atypical network). Several tiny *black dots* were also present. The hypopigmented part of the lesion was characterized mainly by the presence of dotted vessels

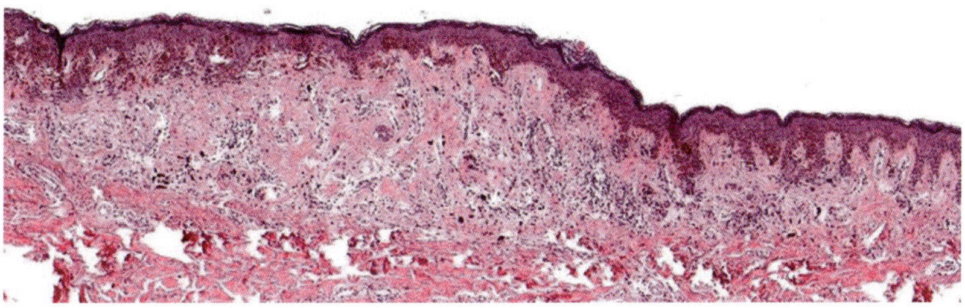

Fig. 4.19 The lesion was quite asymmetrical; the superficial dermal fibrosis was apparent at low-power view

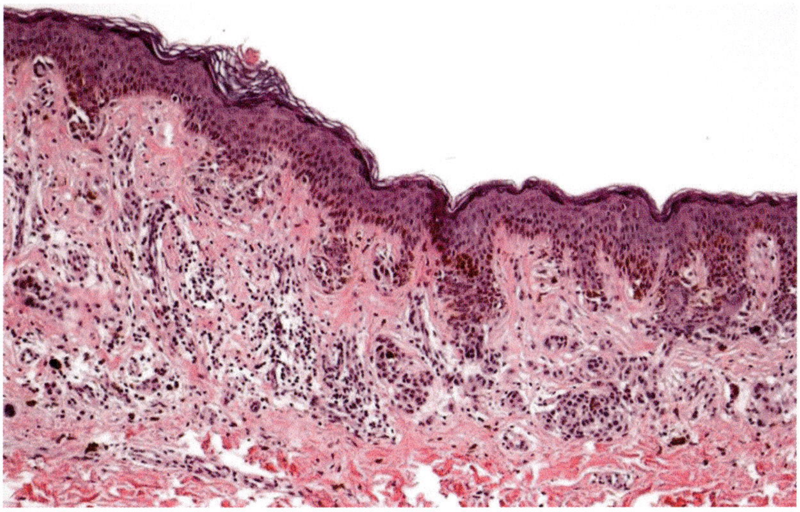

Fig. 4.20 At higher magnification, the lesion had a florid junctional component, arranged in single units and small nests

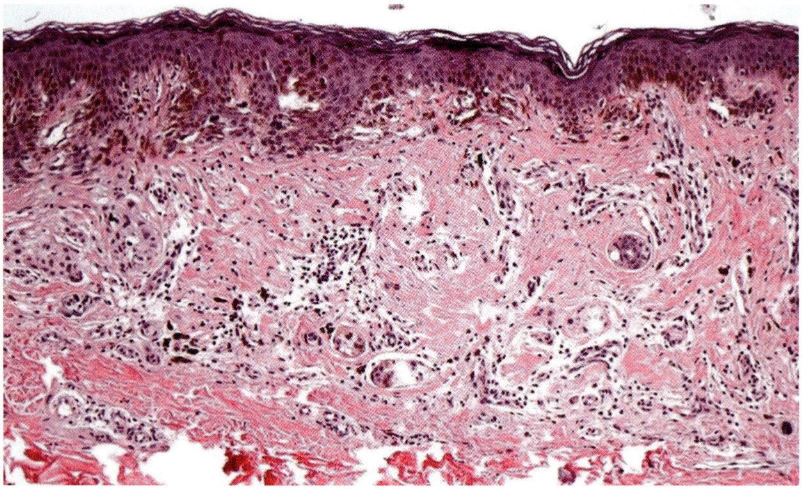

Fig. 4.21 Some small melanocytic nests were embedded in the derma; this showed diffuse scarring, some lymphocytes, and vertically oriented vessels

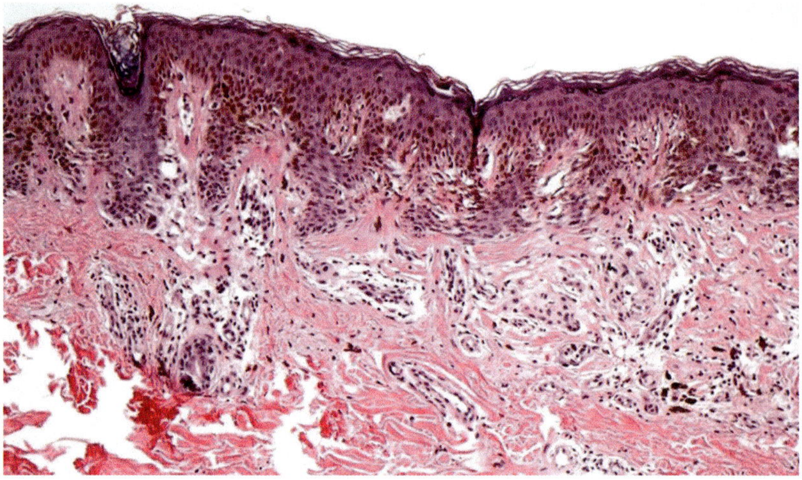

Fig. 4.22 The junctional nests have features reminiscent of Clark nevi

Case 5

Patient: Female, 29 years old

Anatomic site of the lesion: Upper back

History of the lesion: Unknown

Management: The lesion was highly asymmetric. The overall dermoscopic aspect was similar to the one reported to typify recurrent nevi. However, no excision was ever performed, according to the history of the patient, and no scar was evident. The lesion was excised to rule out melanoma.

Histopathologic diagnosis: Compound nevus with dermal sclerosis. A recurrent nevus could not be ruled out.

Key message: Sclerosing nevus with pseudomelanomatous features is known to display regression-like fibrosis. The proliferation of nevus cells on the sclerotic background might occasionally result in an appearance similar to that of nevi recurring after surgical excision. Undoubtedly, this diagnosis can be established only histopathologically and can be accepted in the context of a young individual.

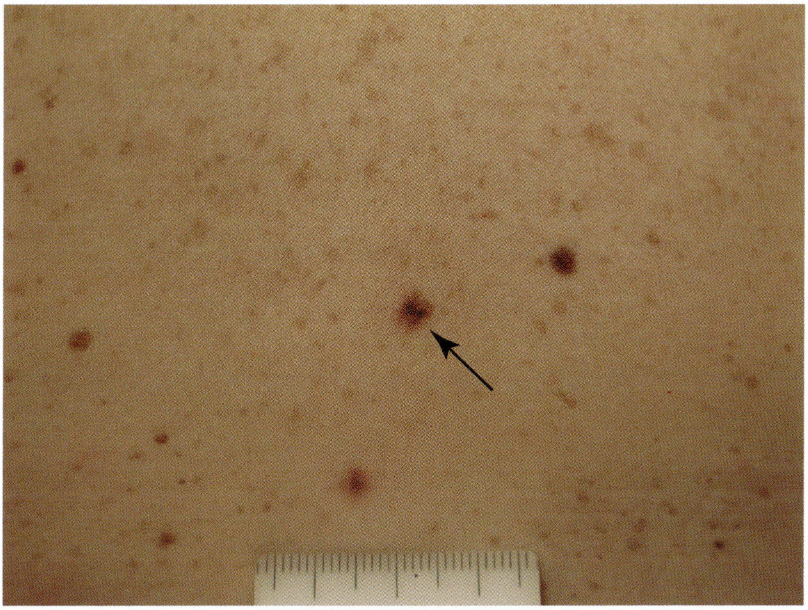

Fig. 4.23 Among multiple nevi on the trunk of the patient, there was a peculiar mole on the upper back, characterized by a dark hyperpigmented center, presence of multiple colors, and a very irregular border

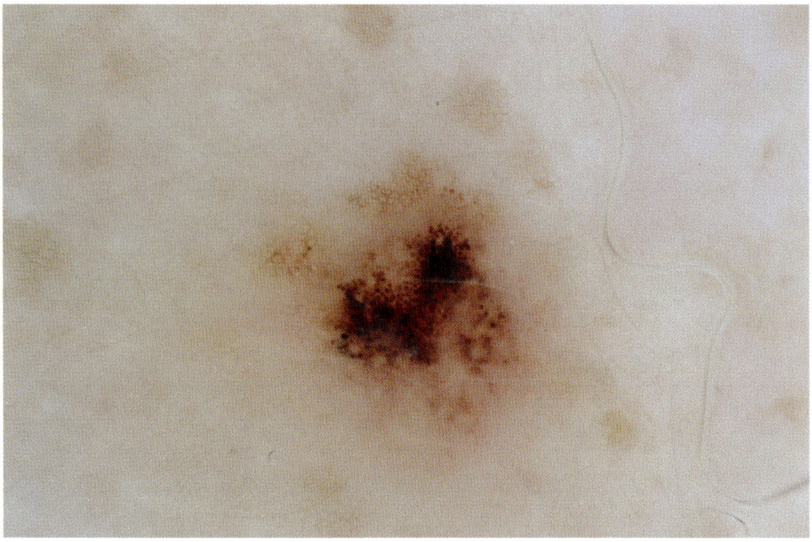

Fig. 4.24 Dermoscopically, the lesion displayed multiple variably sized and distributed *dark-brown* globules in the central part and a *light-brown* network at the periphery

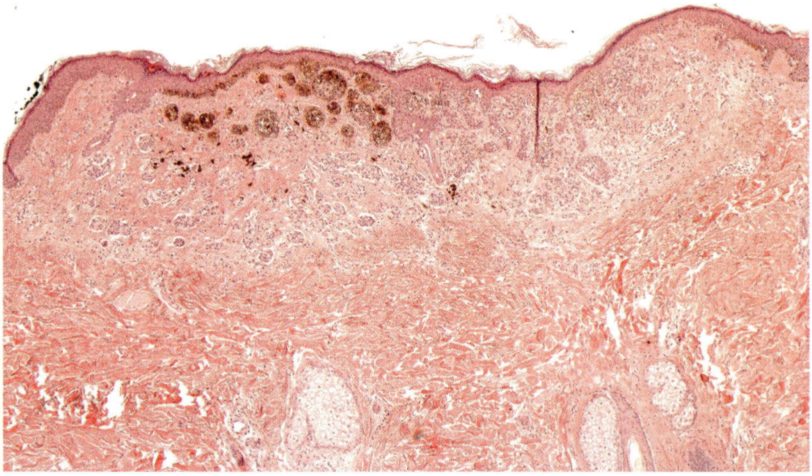

Fig. 4.25 At low-power view, the lesion was asymmetric and irregularly pigmented, mainly arranged in nests

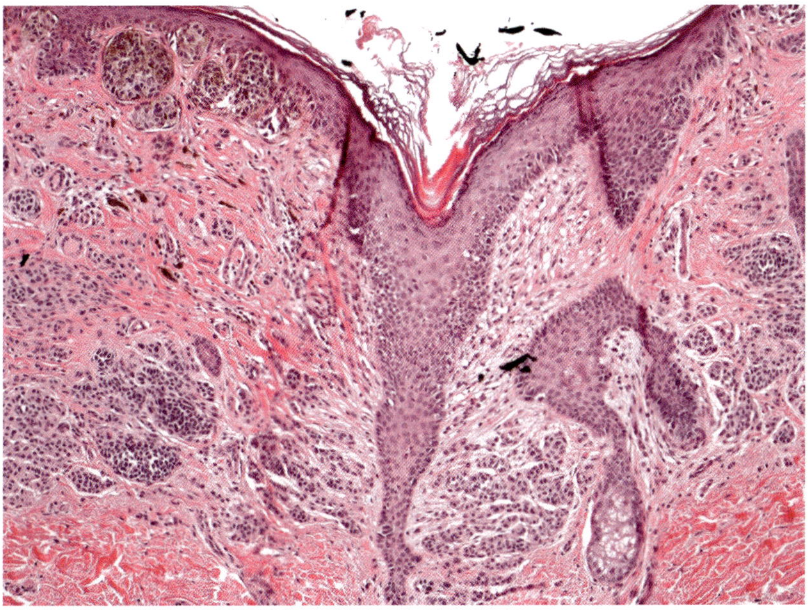

Fig. 4.26 At higher magnification, dermal nests were small, sometimes hyperpigmented, and embedded in a diffuse fibrosis. In the epidermis there was a uniform basal hyperpigmentation

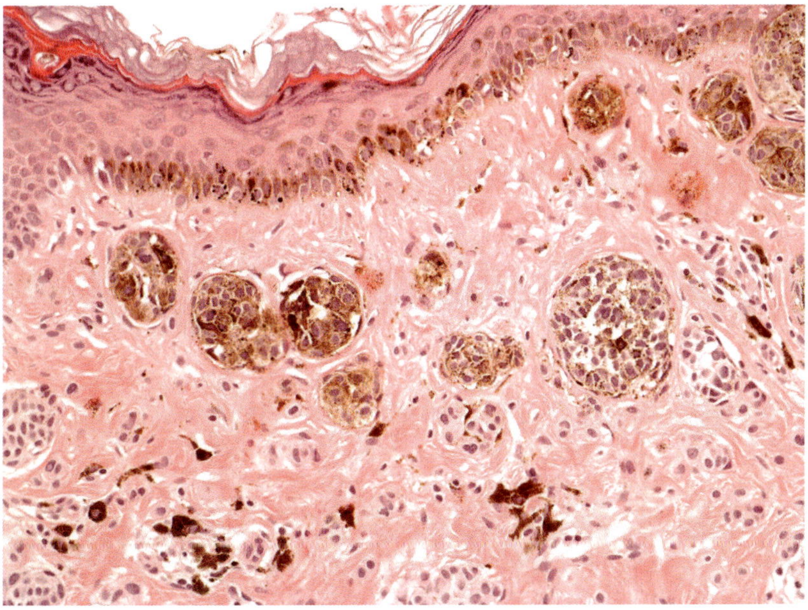

Fig. 4.27 Cytologically, melanocytes were bland and monomorphous

Suggested Reading

1. Blum A, Hofmann-Wellenhof R, Marghoob AA, Argenziano G, Cabo H, Carrera C, Costa Soares de Sá B, Ehrsam E, González R, Malvehy J, Manganoni AM, Puig S, Simionescu O, Tanaka M, Thomas L, Tromme I, Zalaudek I, Kittler H. Recurrent melanocytic nevi and melanomas in dermoscopy: results of a multicenter study of the International Dermoscopy society. JAMA Dermatol. 2014;150(2):138–45.
2. King R, Hayzen BA, Page RN, Googe PB, Zeagler D, Mihm MC Jr. Recurrent nevus phenomenon: a clinicopathologic study of 357 cases and histologic comparison with melanoma with regression. Mod Pathol. 2009;22(5):611–7.
3. Weyant GW, Chung CG, Helm KF. Halo nevus: review of the literature and clinicopathologic findings. Int J Dermatol. 2015;54(10):e433–5.
4. Suzuki N, Dalapicola M, Argenziano G, et al. Halo and pseudo-halo melanoma. J Am Acad Dermatol. 2016;74:e59–61.
5. Ko CJ, Bolognia JL, Glusac EJ. "Clark/dysplastic" nevi with florid fibroplasia associated with pseudomelanomatous features. J Am Acad Dermatol. 2011;64(2):346–51.
6. Ferrara G, Giorgio CM, Zalaudek I, et al. Sclerosing nevus with pseudomelanomatous features (nevus with regression-like fibrosis): clinical and dermoscopic features of a recently characterized histopathologic entity. Dermatology. 2009;219:202–8.
7. Lallas A, Zalaudek I, Apalla Z, et al. Management rules to detect melanoma. Dermatology. 2013;226:52–60.
8. Longo C, Piana S, Lallas A, et al. Routine clinical-pathologic correlation of pigmented skin tumors can influence patient management. PLoS One. 2015;10:e0136031.
9. Fabrizi G, Pennacchia I, Pagliarello C, Massi G. Sclerosing nevus with pseudomelanomatous features. J Cutan Pathol. 2008;35(11):995–1002.
10. Ferrara G, Amantea A, Argenziano G, Broganelli P, Cesinaro AM, Donati P, Pellacani G, Zalaudek I, Tomasini C. Sclerosing nevus with pseudomelanomatous features and regressing melanoma with nevoid features. J Cutan Pathol. 2009;36(8):913–5.

5 Spitzoid Lesions

Spitzoid-looking lesions represent a fascinating group of tumors, regarding their morphology and growth dynamic. Management of spitzoid-looking tumors is controversial, mainly because of a known morphologic overlap between Spitz nevi and spitzoid melanoma.

Spitz nevi are known to display peculiar and characteristic dermoscopic patterns. The most frequent and well-known one is the so-called "starburst" pattern which typifies pigmented Spitz (Reed) nevi. The "starburst" pattern consists of a darkly pigmented black (or bluish) central area and peripheral symmetrically distributed streaks or pseudopods. Another frequent dermoscopic pattern of pigmented Spitz nevi consists of multiple pigmented globules surrounded by depigmented (white) network-like lines (reticular depigmentation). The latter pattern is also called "inverse network," resulting from white (depigmented) lines and pigmented holes (corresponding to the globules). Nonpigmented Spitz nevi are dermoscopically typified by the presence of multiple and, usually, symmetrically distributed dotted vessels, typically associated with a network of white lines very similar to the one described above for pigmented Spitz nevi. In fact, the vessels of flat/elevated nonpigmented Spitz nevi are dotted; in nodular nevi they often project as larger red globules or glomerular (coiled) or even tortuous vessels.

All patterns described above to typify Spitz nevi are characterized by a morphologic symmetry of colors and structures. An asymmetric distribution of the aforementioned features (streaks, pseudopods, vessels) should not be interpreted as suggestive of a Spitz nevus, since all these criteria might also be seen in melanoma.

Spitz nevi not displaying a dermoscopic symmetry do exist, but these nevi cannot be safely recognized as such and should be excised to rule out melanoma. Spitz nevi with peculiar dermoscopic features have also been described and possibly correspond to specific histopathologic variants. Again, the lesions cannot be safely classified as Spitz nevi on the basis of clinical and dermoscopic morphology.

In contrast to the morphologic symmetry of Spitz nevi, spitzoid melanoma is typically characterized by a morphologic asymmetry in terms of colors and/or structures. Features seen in Spitz nevi might also be found in melanoma, but their distribution is usually not symmetric. Asymmetrically distributed peripheral streaks or pseudopods or asymmetrically distributed dotted vessels represent well-known melanoma-specific criteria. Therefore, any lesion dermoscopically typified by a morphologic asymmetry and the presence of these criteria should be excised to rule out melanoma, irrespectively of the patient's age.

The real controversy in the management of spitzoid-looking lesions originates from the fact that, rarely, spitzoid melanoma might lack its usual asymmetry and display a perfectly

C. Longo et al., *Atlas of Diagnostically Challenging Melanocytic Neoplasms*,
https://doi.org/10.1007/978-3-319-48653-6_5

symmetric spitzoid pattern (starburst or globular with inverse network or dotted vessels with inverse network). Therefore, these melanomas might be missed resulting in undesirable consequences. To address these problems, two main strategies have been developed: the most conservative strategy suggests excision of any spitzoid-looking lesion (symmetric or asymmetric) irrespectively of the age of the patient. Obviously, this strategy would ensure the minimal risk of missing melanoma but would also result in a high number of unnecessary excisions of benign tumors (Spitz nevi). The latter consequence could be quite problematic, since the majority of Spitz nevi develop in young children, and their excision might require general anesthesia. To minimize this limitation, a more flexible age-dependent strategy has been advised, based on evidence demonstrating that the likelihood of a perfectly symmetric spitzoid-looking lesion to be a melanoma strongly depends on the age of the patient. The latest recommendations include the following:

1. All asymmetric lesions with spitzoid features should be excised.
2. All symmetric spitzoid-looking nodules should be excised.
3. The management of symmetric flat spitzoid-looking lesions should depend on the age: Below 12 years of age, these lesions might be monitored until stabilization. In individuals older than 12 years, they should be excised because the possibility of melanoma becomes considerable.

Case 1

Patient: Female, 38 years old

Anatomic site of the lesion: Right buttock

History of the lesion: Unknown

Management: The dermoscopic pattern was not very characteristic. The presence of multiple globules and dotted vessels was highly suggestive of a melanocytic lesion, possibly spitzoid. The lesion was excised to rule out melanoma.

Histopathologic diagnosis: Compound nevus with a spitzoid junctional and a usual dermal component (so-called combined nevus)

Key message: In order to detect early melanomas, the presence of one melanoma-specific criterion is enough to warrant excision of the lesion. In the present case, irregularly distributed globules were evident. The simultaneous presence of evident dotted vessels raised also the suspicion of a spitzoid lesion. Spitzoid-looking lesions in adults should also be excised, since the possibility of melanoma cannot be excluded. In the fourth decade of life, the possibility of a spitzoid-looking lesion to be a Spitz nevus is still higher than that of melanoma. Consequently, the histopathologic diagnosis of compound nevus with a spitzoid component is acceptable, since it is compatible with the morphology and the age of the patient.

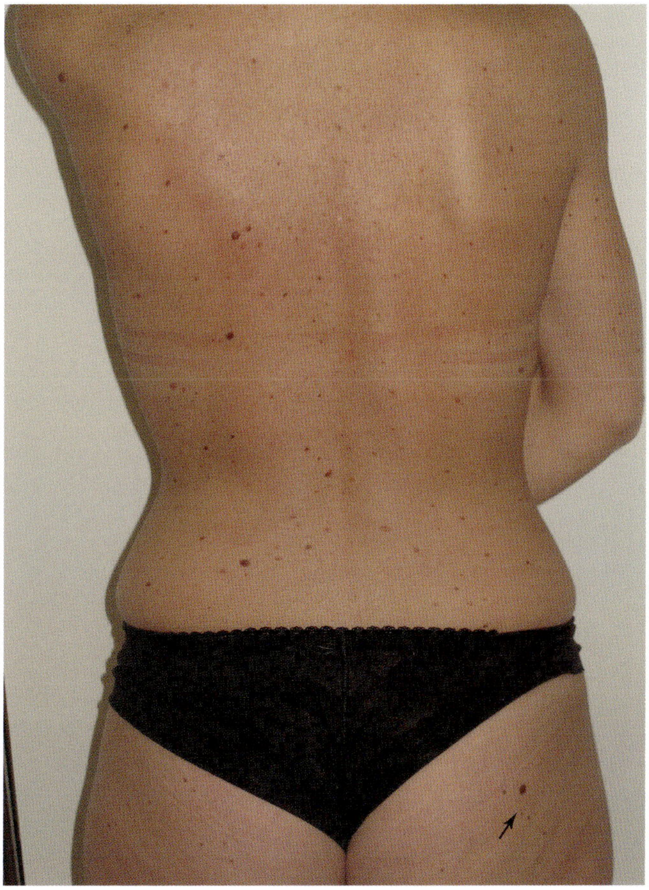

Fig. 5.1 The patient had several moles on the trunk and the extremities. The macroscopic overview did not reveal any peculiar mole

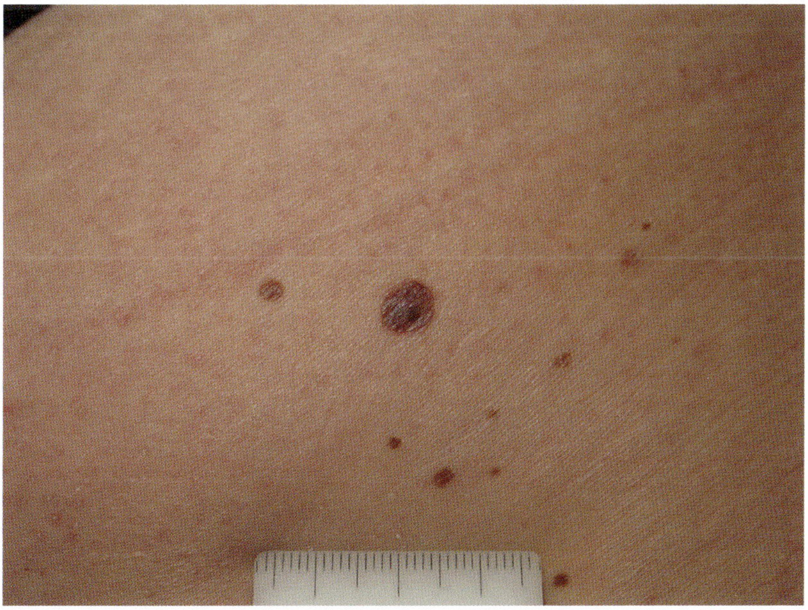

Fig. 5.2 Close-up clinical examination of the largest moles on the right buttock revealed a roundish brown-colored lesion with focal central hyperpigmentation

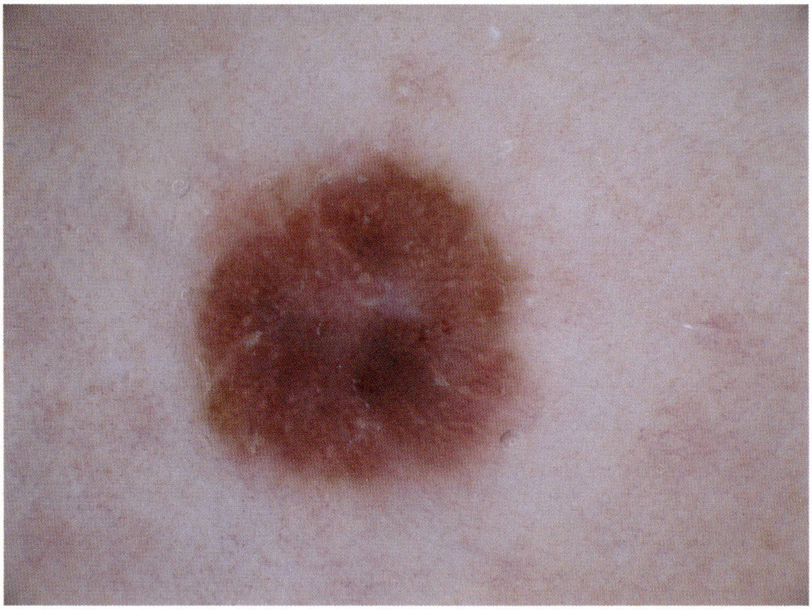

Fig. 5.3 Dermoscopically, the lesion was characterized by multiple and irregularly distributed dark-brown-to-black globules and multiple dotted vessels

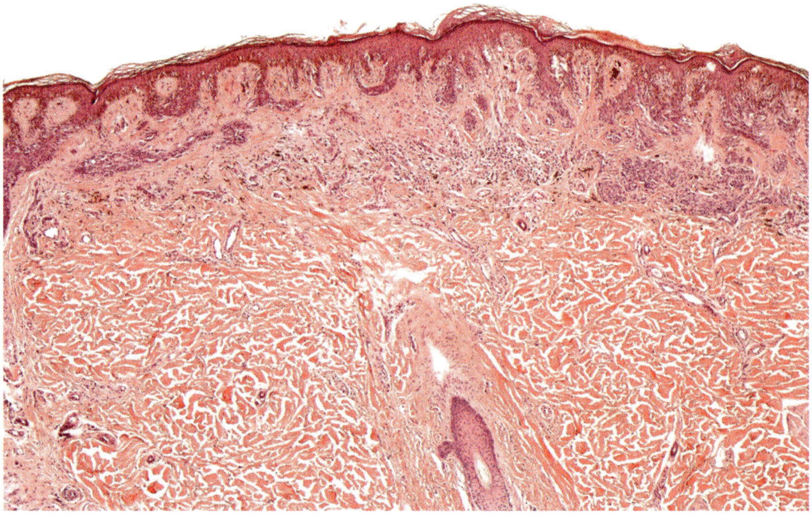

Fig. 5.4 On histology, an asymmetrical melanocytic proliferation was present, with a prominent junctional activity. Pigmentation was abundant

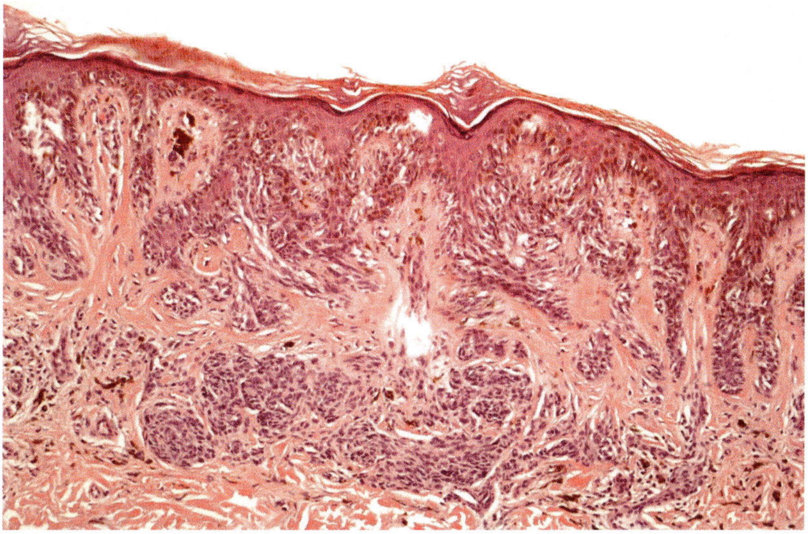

Fig. 5.5 Melanocytes at the junction had a spitzoid appearance, while the dermal component was a usual-looking benign nevus

Case 2

Patient: Female, 42 years old

Anatomic site of the lesion: Left lower leg

History of the lesion: Appeared 2 years ago and gradually enlarged

Management: Because of the asymmetry of colors and structures at the right part of the lesion and because of the age of the patient, the lesion was excised to rule out melanoma.

Histopathologic diagnosis: Spitz/Reed nevus

Key message: Spitzoid-looking lesions are known to represent a diagnostically challenging group of tumors. The controversy exists because spitzoid melanoma might dermoscopically display a perfectly symmetric spitzoid pattern. To address this problem, an age-dependent management strategy should be applied. According to it, spitzoid-looking lesions in patients older than 12 years should be excised to rule out melanoma. Until the age of 50 years, more than 50% of these lesions will still prove to be Spitz nevi, as was the present case.

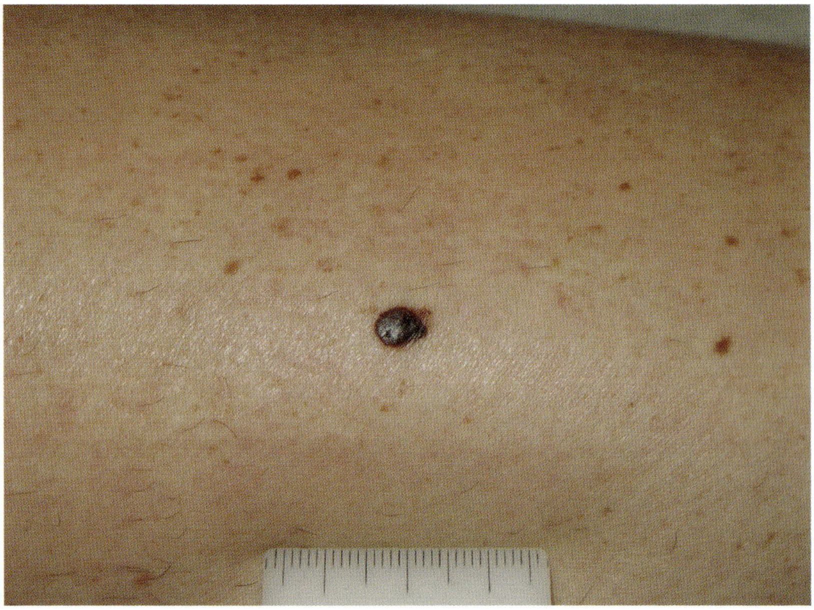

Fig. 5.6 An elevated roundish papule of dark-brown color with a diameter of 5 mm. At the upper right part of the lesion, flat component was present, resulting in an overall-asymmetric shape

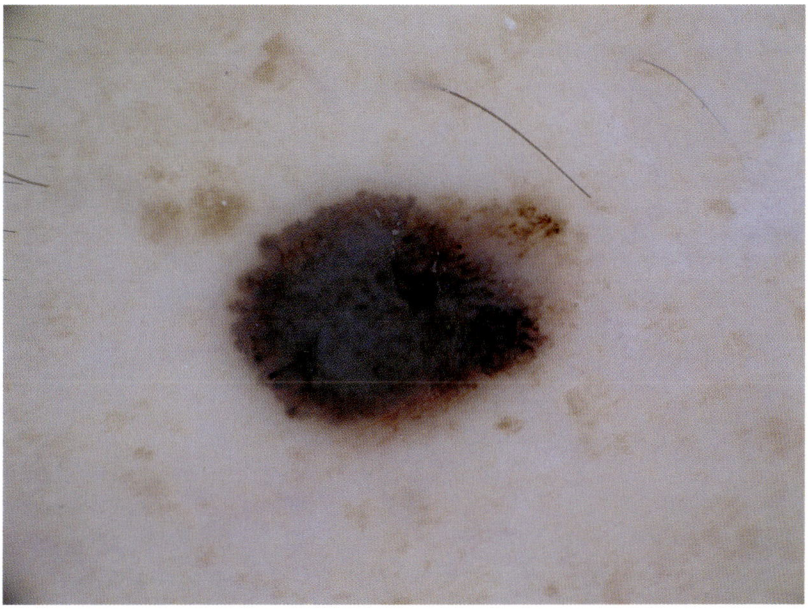

Fig. 5.7 Dermoscopy revealed a blue-black homogeneous central part and multiple peripheral streaks and pseudopods. This dermoscopic pattern is well-known to be suggestive of pigmented Spitz (Reed) nevus

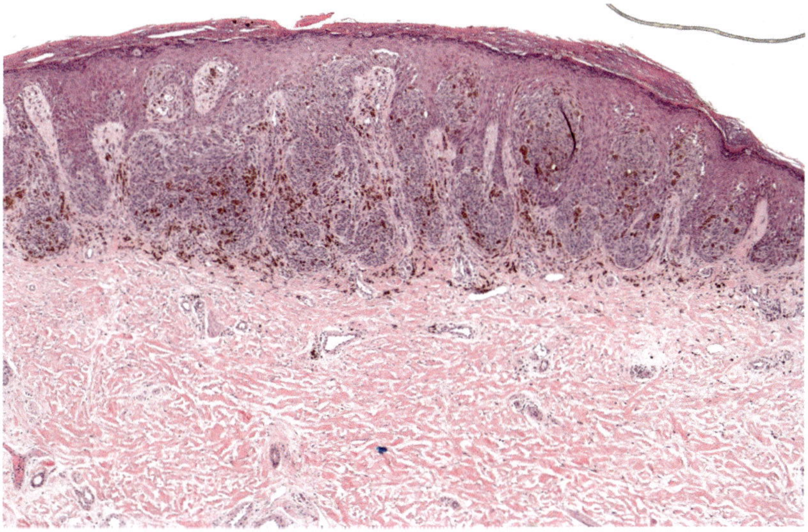

Fig. 5.8 Histologically, the lesion was a rather symmetrical proliferation of spindle, pigmented melanocytes, mainly located at the tip of the epidermal papillae

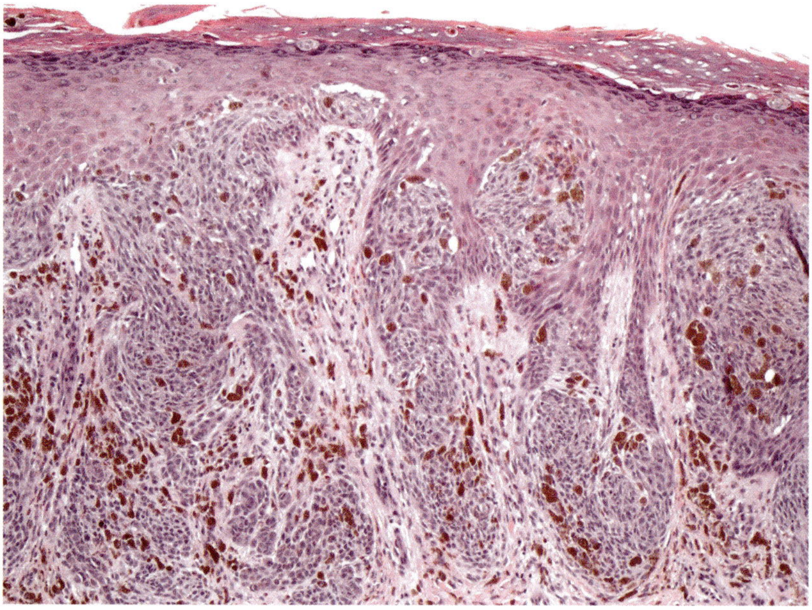

Fig. 5.9 Melanocytes were quite monomorphous and basally located. The melanocytic ascent to the superficial epidermal layers was rare. Heavily pigmented melanophages were present in the superficial dermis

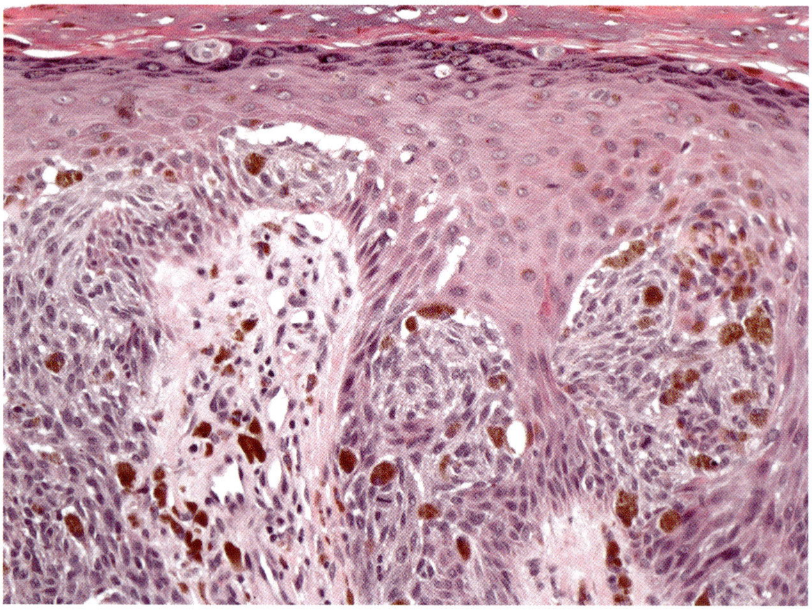

Fig. 5.10 Pigmentation was abundant and was also seen in the stratum corneum

Case 3

Patient: Female, 26 years old

Anatomic site of the lesion: Right thigh

History of the lesion: The lesion appeared 1 year ago and gradually enlarged.

Management: The differential diagnosis included a spitzoid lesion and a vascular tumor. The reddish roundish area was not perfectly homogeneous and sharply demarcated, and vascular structures could be seen. The latter, along with the presence of inverse network, was not compatible with the diagnosis of angioma, which prompted us to excise the lesion to rule out spitzoid melanoma.

First histopathologic diagnosis: Compound epithelioid Spitz nevus

Key message: The typical dermoscopic pattern of nonpigmented Spitz nevus consists of multiple and regularly distributed dotted vessels, often combined with the so-called inverse network, which corresponds to a network of white intermingled lines surrounding the vascular structures. In nodular Spitz nevi, the vascular structures might project not as dots, but as larger red globules or even coiled or tortuous vessels. Nodular spitzoid-looking lesions should be always excised because the possibility of atypical Spitz tumor or spitzoid melanoma cannot be excluded. In this case, the tumor was shown to be benign (Spitz nevus).

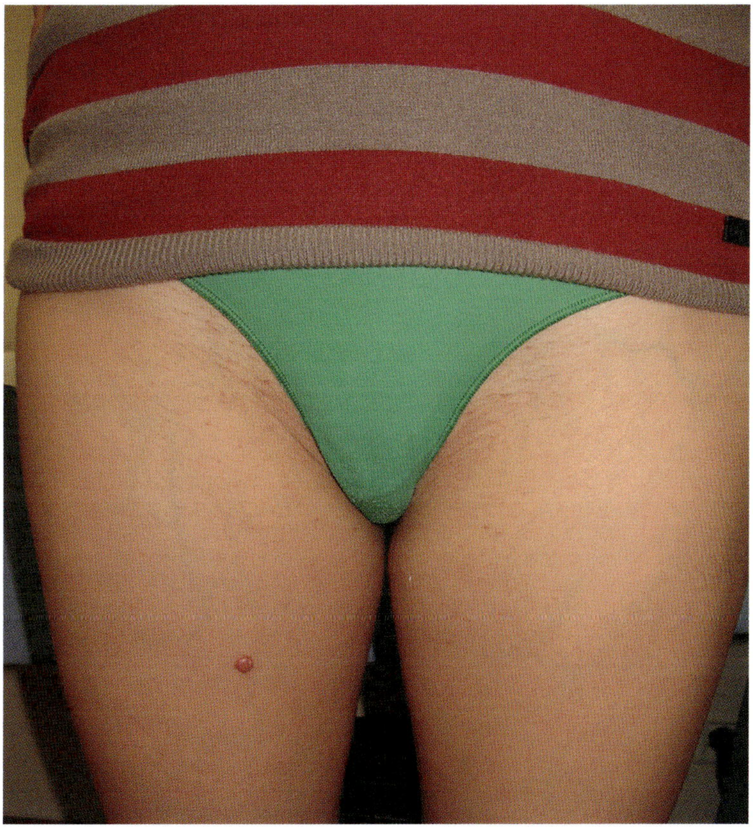

Fig. 5.11 A nonpigmented nodule on the thigh

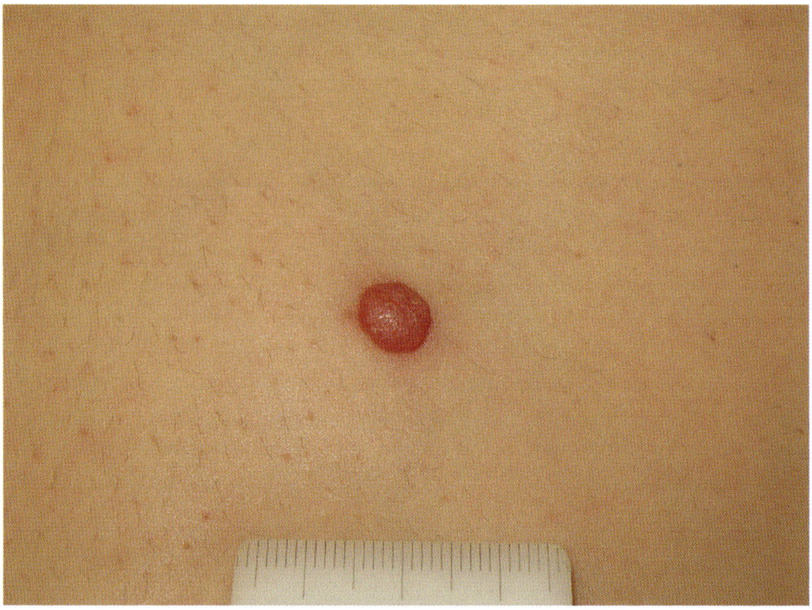

Fig. 5.12 Clinical close-up examination revealed a symmetric roundish nodule without evidence of pigmentation

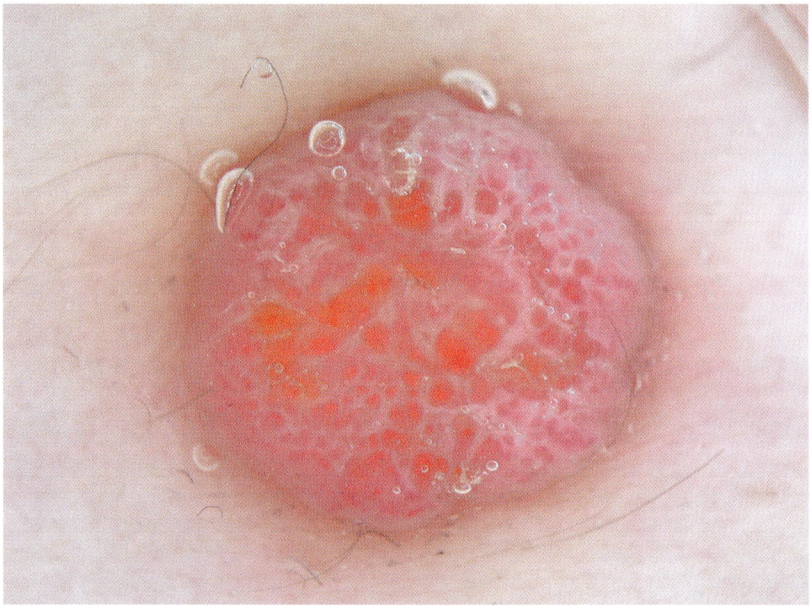

Fig. 5.13 Dermoscopy revealed multiple roundish red areas, each one surrounded by a whitish halo. The lines of neighboring halos were connected to each other so as to form a white network, corresponding to the so-called reticular depigmentation or inverse network

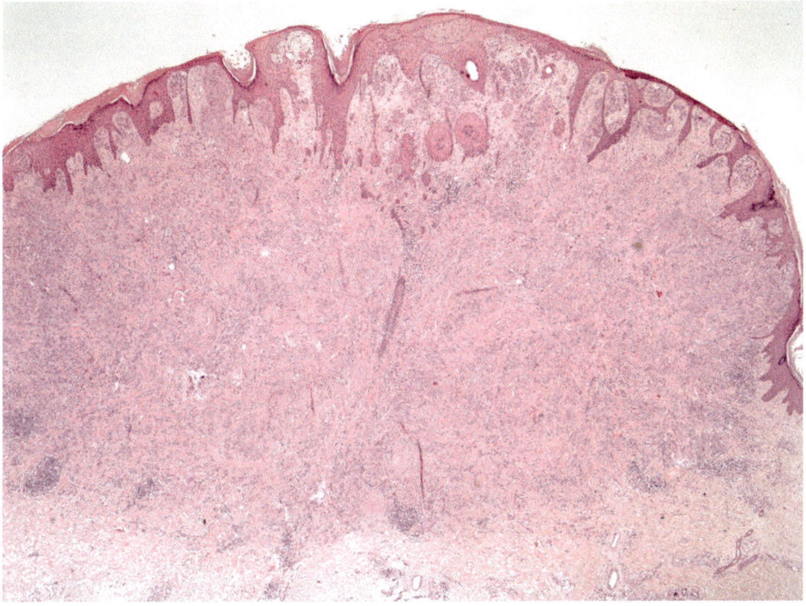

Fig. 5.14 On histology, the lesion was rather symmetrical and vaguely wedge shaped. The overlying epidermis was irregularly hyperplastic

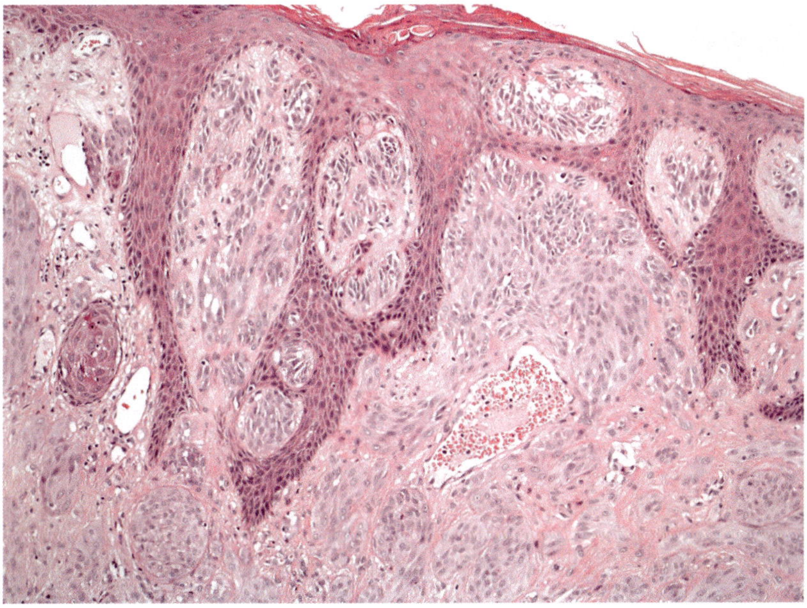

Fig. 5.15 The lesion was composed of confluent nests of epithelioid, spitzoid cells. Mitotic figures were negligible

Case 4

Patient: Male, 54 years old

Anatomic site of the lesion: Left upper arm

History of the lesion: The patient had recently noticed this slightly pigmented papule on the left elbow, which appeared few months before and gradually enlarged.

Management: The lesion was excised to rule out melanoma.

Histopathologic diagnosis: Spitzoid melanoma

Key message: The combination of white lines and dotted vessels is typical of Spitz nevi in dermoscopy. However, melanoma may be dermoscopically indistinguishable from Spitz nevi, exhibiting a symmetric pattern mimicking Spitz nevus. The only safe strategy not to miss melanoma is to excise all spitzoid-looking lesions in adults.

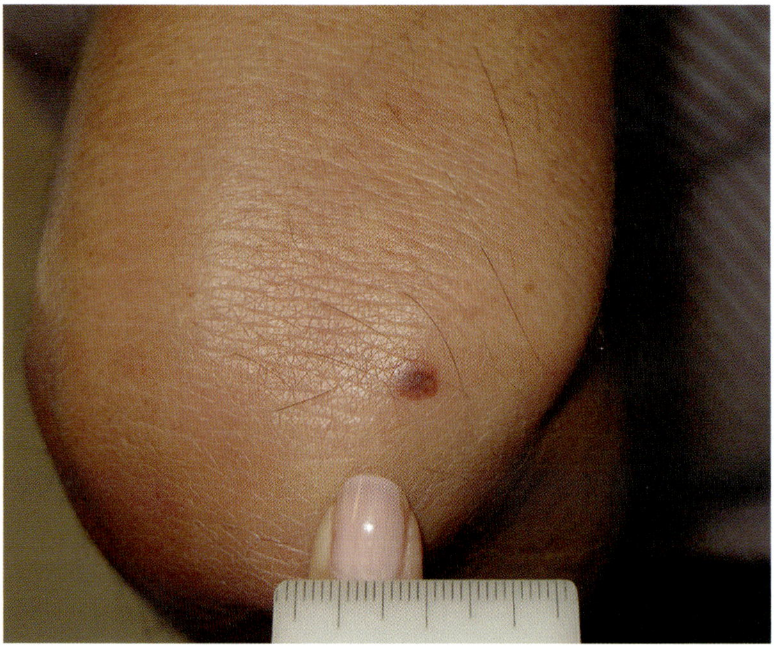

Fig. 5.16 Close-up clinical examination revealed a roundish light-brown lesion 0.5 cm in diameter, with regular borders and slightly hypopigmented periphery

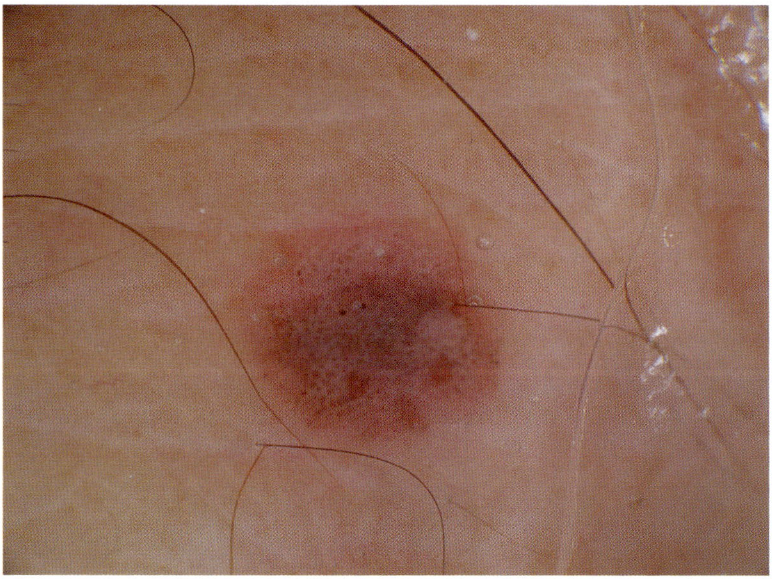

Fig. 5.17 Dermoscopically, the lesion was characterized by dotted vessels intersected by white lines. A bluish area was visible in the center of the lesion that appeared overall symmetric

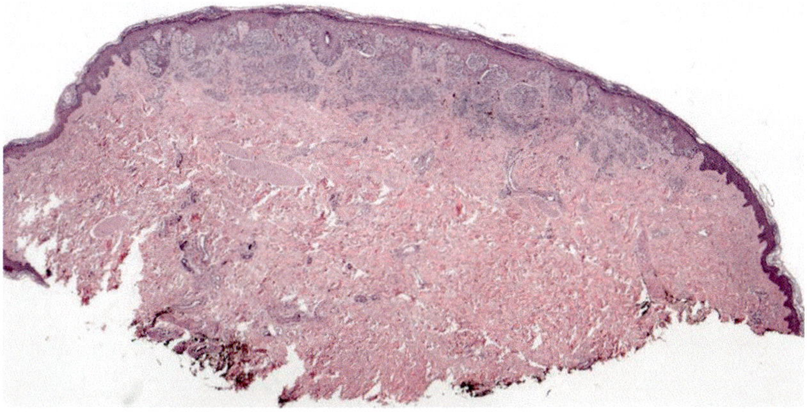

Fig. 5.18 At low-power view, the lesion was a symmetrical, hypopigmented papule

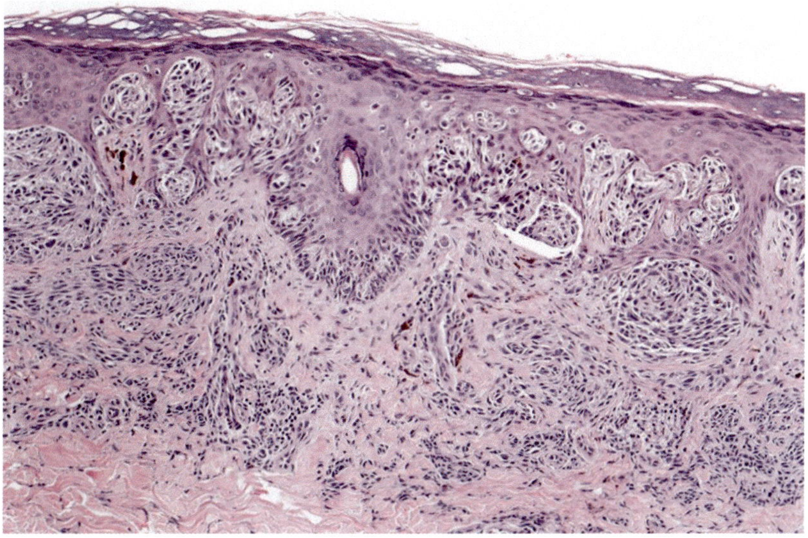

Fig. 5.19 At higher magnification, the lesion was constituted by atypical melanocytes with striking upward migration through the epidermis

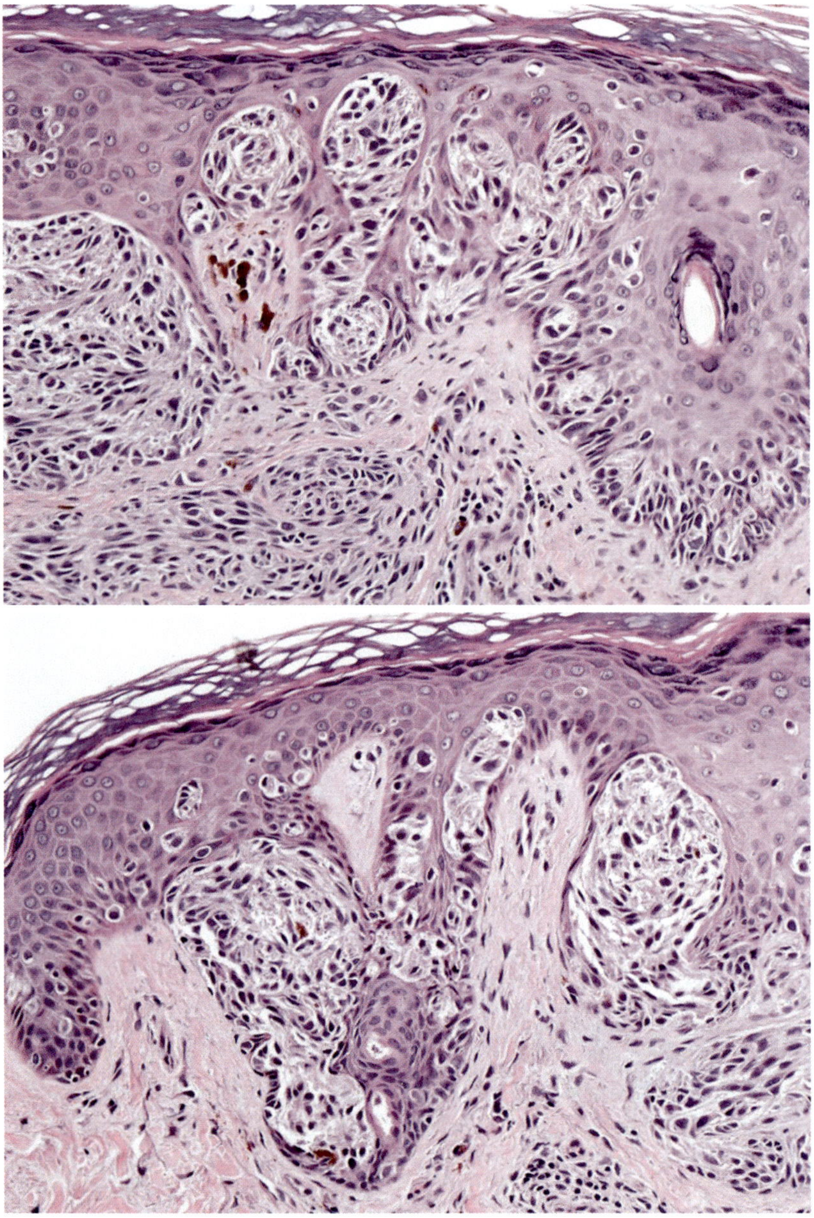

Figs. 5.20 and 5.21 Cytologically, neoplastic cells had hyperchromatic nuclei and abundant cytoplasm. The infiltrating attitude toward the epidermis was diffuse

Case 5

Patient: Male, 26 years old

Anatomic site of the lesion: Left lower limb

History of the lesion: The patient had several moles on the trunk and the extremities. Among them, there was this newly appearing lesion on the left thigh.

Management: The lesion was excised to rule out melanoma.

Histopathologic diagnosis: Compound Spitz nevus

Key message: This case is very similar to case 4 both dermoscopically and histologically. The combination of white lines and dotted vessels is typical of Spitz nevi in dermoscopy. However, since melanoma may be dermoscopically indistinguishable from Spitz nevi, all spitzoid-looking lesions in adults should be excised.

Histological differential diagnosis between a Spitz nevus and a spitzoid melanoma can be challenging, as in the present case; a consistently minor degree of epidermal infiltration in comparison with case 4 is the major criterion for differentiating these entities in the present cases.

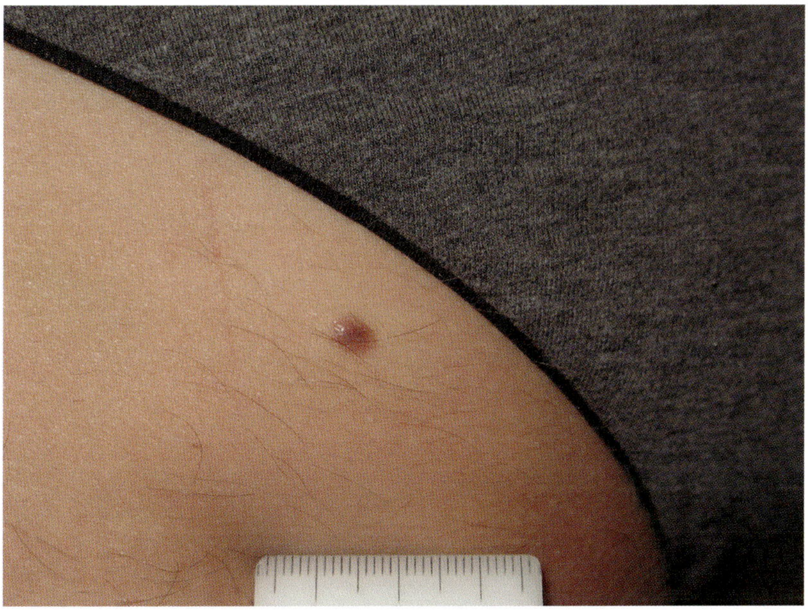

Fig. 5.22 Close-up clinical examination revealed a roundish light-brown-to-pink papule lesion with a diameter of 0.5 cm, with regular borders

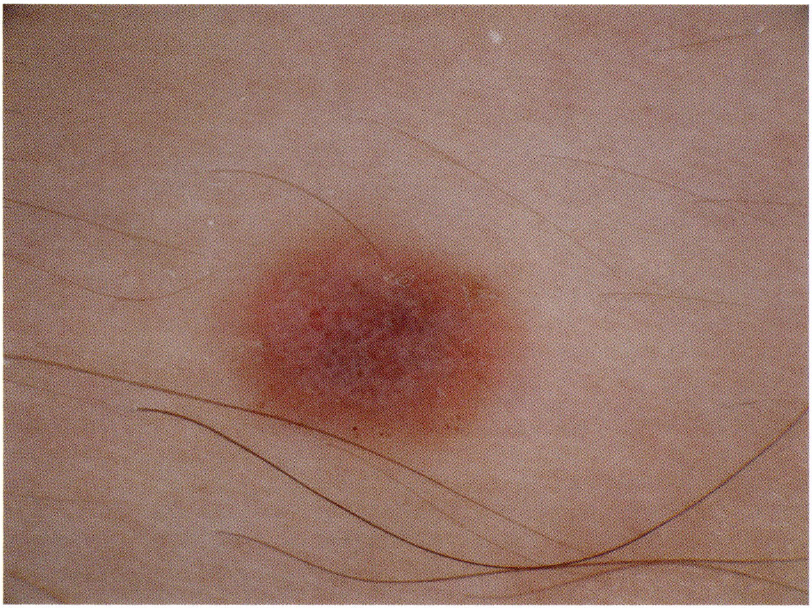

Fig. 5.23 Dermoscopically, the lesion was characterized by dotted vessels intersected by white lines. A bluish area was visible in the center of the lesion that was overall symmetric

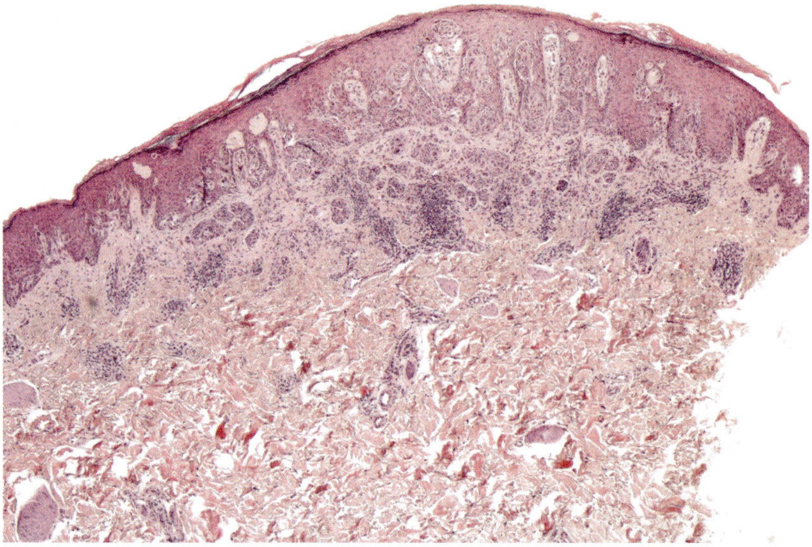

Fig. 5.24 At scanning view, the lesion was dome shaped but not completely symmetrical

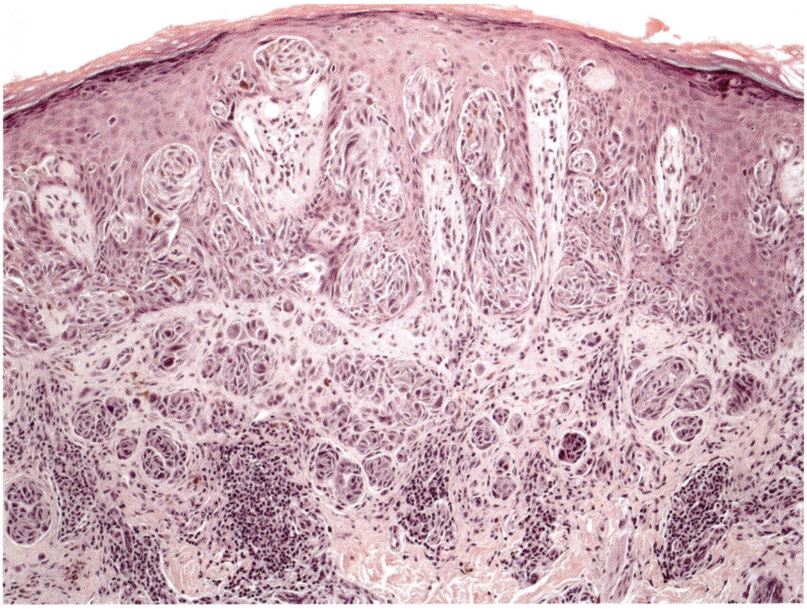

Fig. 5.25 Melanocytes were epithelioid and spindle, arranged in dyscohesive nests with retraction artifact. No mitoses were present in the dermal component

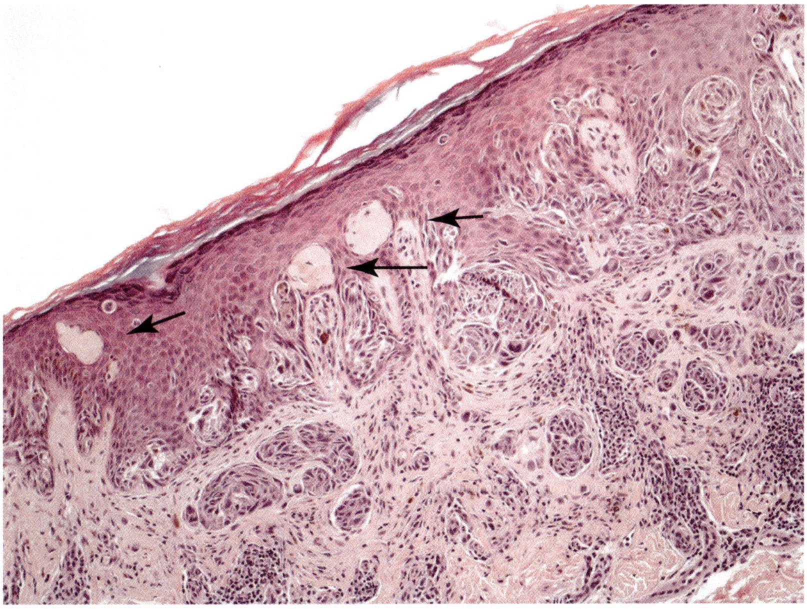

Fig. 5.26 Single foci of pagetoid spread could be seen, but most proliferating cells were restricted to the lower epidermal layers. Kamino bodies are numerous (*arrows*)

Case 6

Patient: Male, 52 years old

Anatomic site of the lesion: Left upper limb

History of the lesion: The lesion appeared few months ago and gradually enlarged.

Management: At the first clinical ground, the lesion was thought to be a dermatofibroma. Looking closer with the dermatoscope, the presence of dotted vessels and inverse network were suggestive for a melanocytic proliferation, and thus the lesion was excised to rule out melanoma.

Histopathologic diagnosis: Atypical spitzoid tumor

Key message: Atypical spitzoid tumors represent a special category of lesions that are difficult to be diagnosed on clinical and histopathologic basis. In this case, the presence of dotted vessels and inverse network were suggestive for a melanocytic atypical proliferation although a clear-cut diagnosis was not possible. However, a prompt surgical excision was made because the lesion was not fulfilling any criteria for benign lesion.

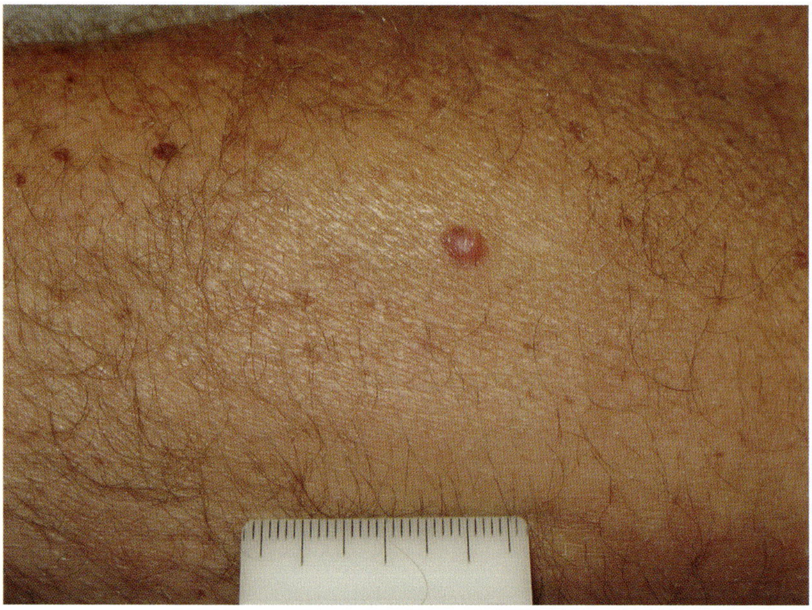

Fig. 5.27 Clinical close-up examination revealed a symmetric roundish papule with a diameter of 4 mm without any remarkable sign of pigmentation

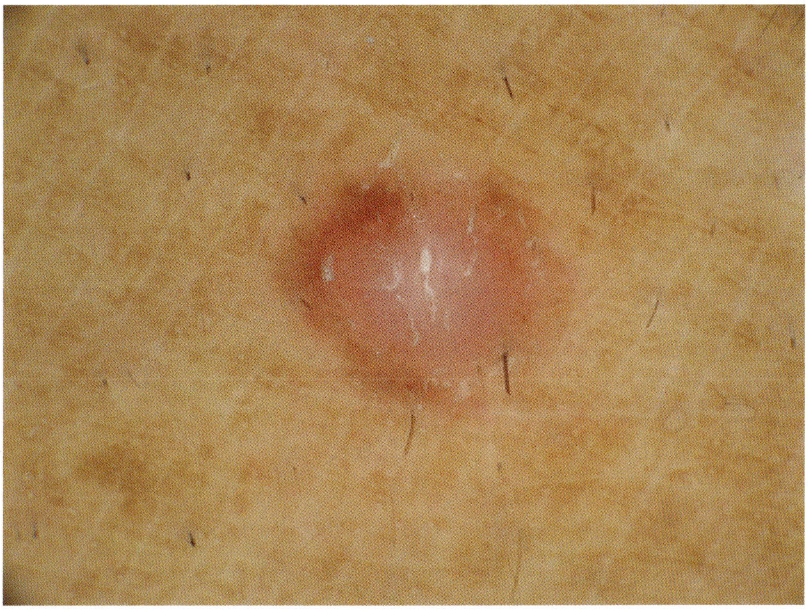

Fig. 5.28 Dermoscopy revealed a pinkish central area with few and barely visible dotted vessels—at the periphery a tiny brown network and an inverse network were noted

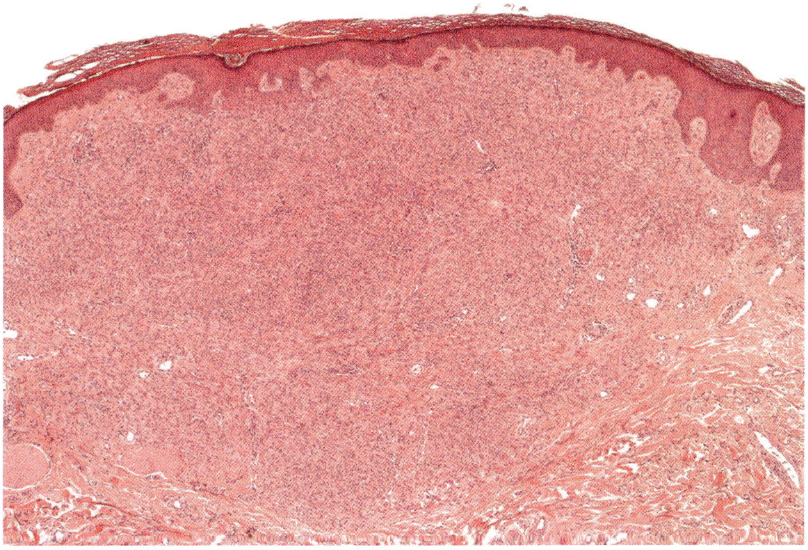

Fig. 5.29 The lesion was nodular, well defined, and dermal based

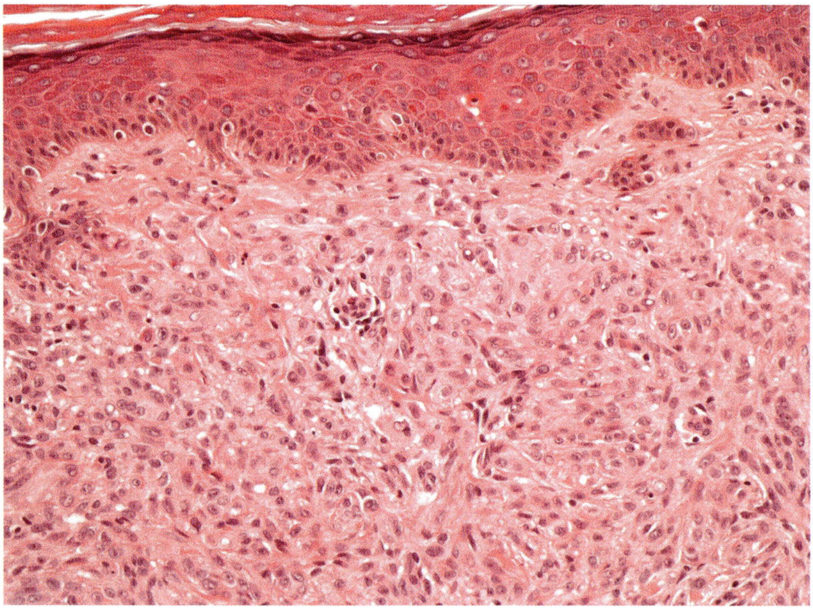

Fig. 5.30 Melanocytes were monomorphous, epithelioid, and growing in solid sheets more than in nests. In the deep dermis, some mitotic figures could be seen

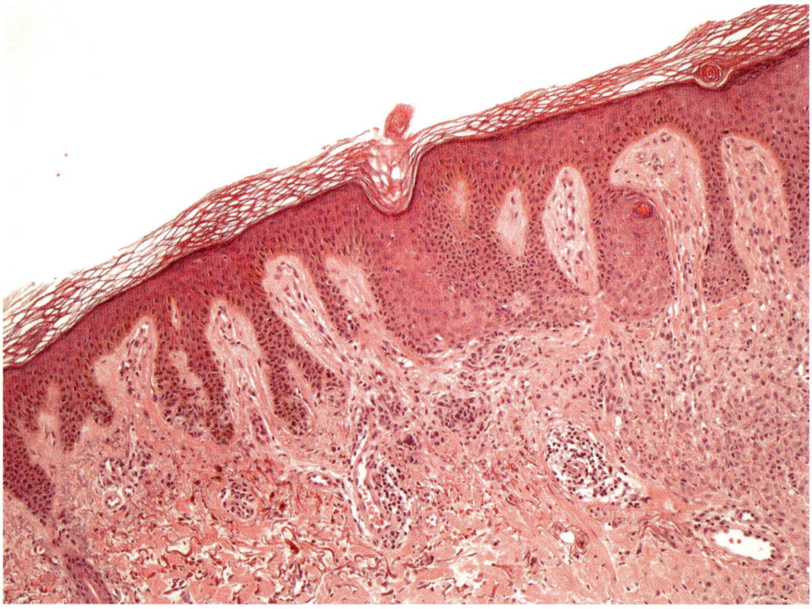

Fig. 5.31 The epidermis was focally hyperplastic but mostly uninvolved by the melanocytic proliferation

Case 7

Patient: Male, 25 years old

Anatomic site of the lesion: Upper limb

History of the lesion: Appeared few months before and gradually enlarged. The patient had several moles on the trunk and the extremities. Among them, there was a relatively large mole on his right arm.

Management: The lesion was excised to rule out melanoma.

Histopathologic diagnosis: Spitz nevus

Key message: One more case highlighting the difficulties in the clinical and dermoscopic differential diagnosis of Spitz nevus and melanoma. Spitz nevi may simulate melanoma exhibiting worrisome clinical and dermoscopic features. Blue-white veil and multicomponent pattern are among the most common melanoma features. Excision was mandatory in this case in order to rule out melanoma.

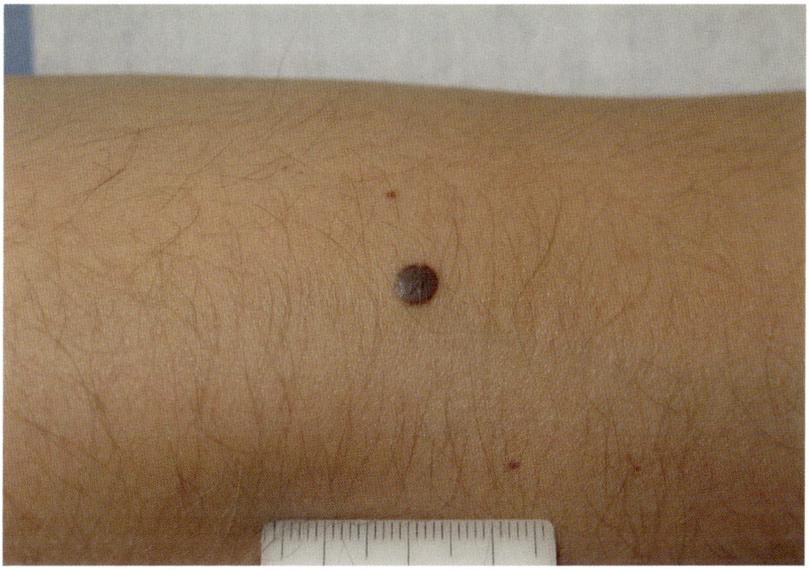

Fig. 5.32 Close-up clinical examination revealed a roundish brown papule with a diameter of 0.7 cm, with a hyperpigmented center and regular borders

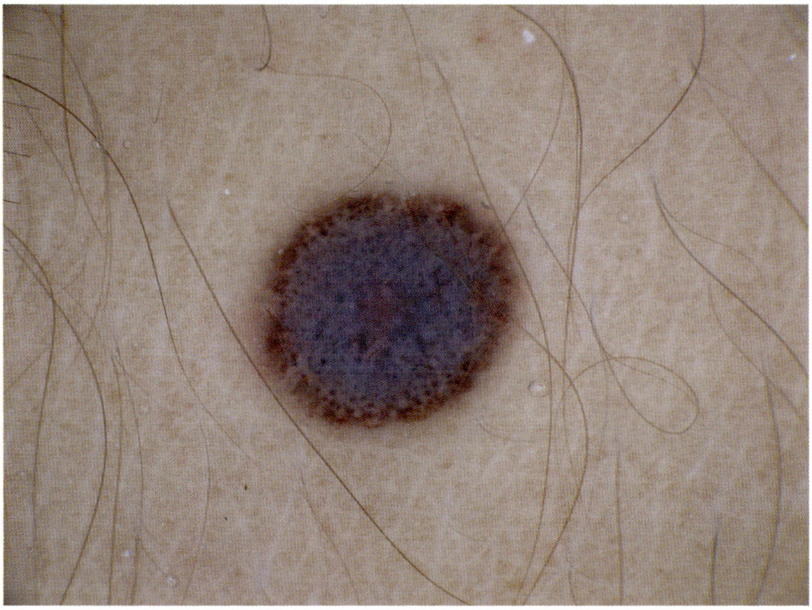

Fig. 5.33 Dermoscopically, the lesion was characterized by a blue-white veil in the center of the lesion and irregularly arranged brown-to-black globules at the periphery

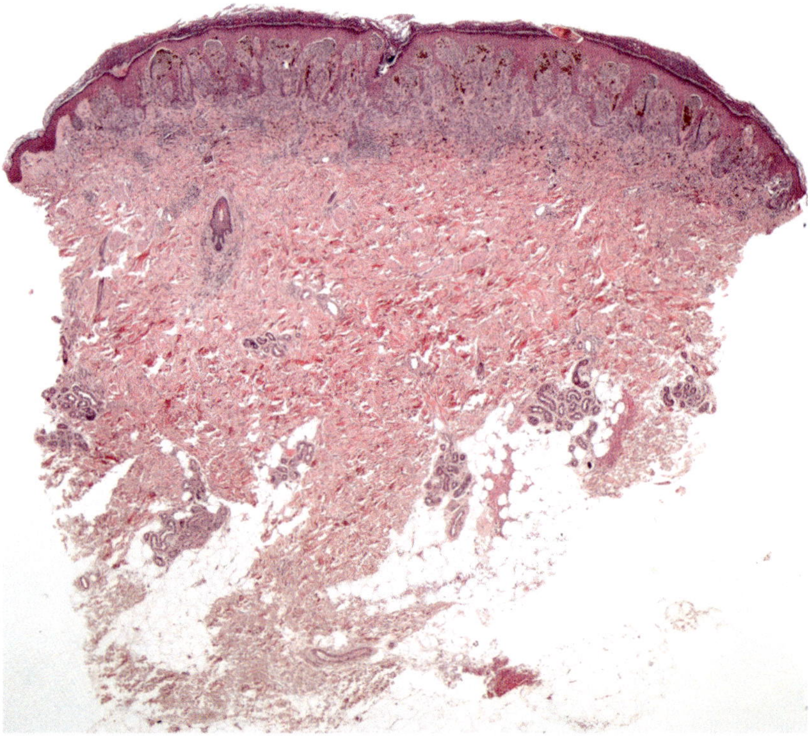

Fig. 5.34 At scanning view, the lesion was a perfectly symmetrical papule

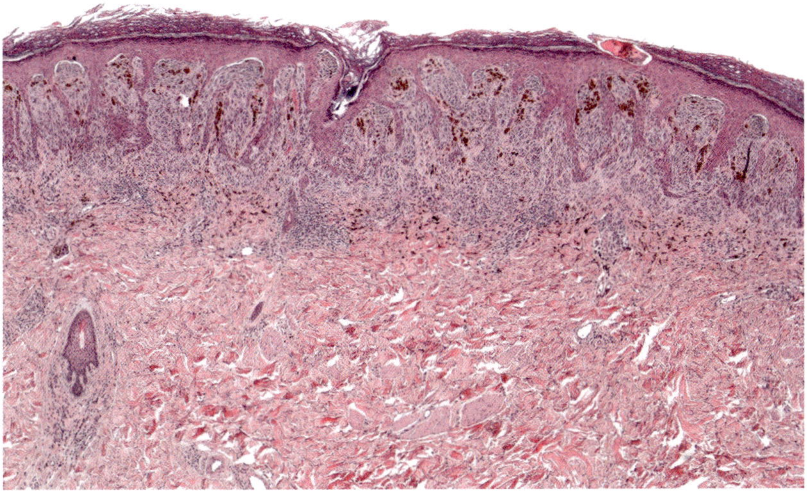

Fig. 5.35 Melanocytes were arranged in vertically oriented nests with typical peripheral clefting

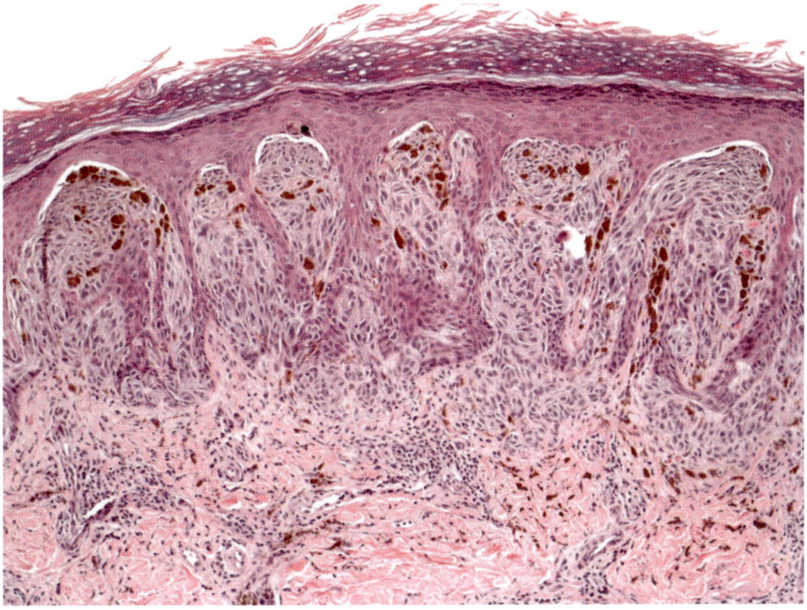

Fig. 5.36 Cytologically, melanocytes were epithelioid and spindle, with coarse pigment granules. The epidermis was hyperplastic and hyperkeratotic

Case 8

Patient: Female, 19 years old

Anatomic site of the lesion: Left lower limb

History of the lesion: The lesion appeared 1 year ago and gradually enlarged.

The body site and the presence of crystalline structures were suggestive for the diagnosis of dermatofibroma. However, the exophytic aspect of the lesion and the presence of atypical vessels could not exclude the possibility of melanocytic lesion such as melanoma or spitzoid tumor.

Management: The lesion was excised to rule out spitzoid melanoma.

Histopathologic diagnosis: Atypical spitzoid tumor

Key message: Atypical Spitz tumors are polymorphic melanocytic proliferations with a nodular clinical appearance. Dermoscopically they demonstrate a multicomponent and nonspecific pattern. Thus, any nodular red lesions with no clear-cut benign dermoscopic features should be excised. Furthermore, a typical nonpigmented spitzoid pattern on dermoscopy (with dotted vessels and white lines) does not exclude atypical Spitz tumors.

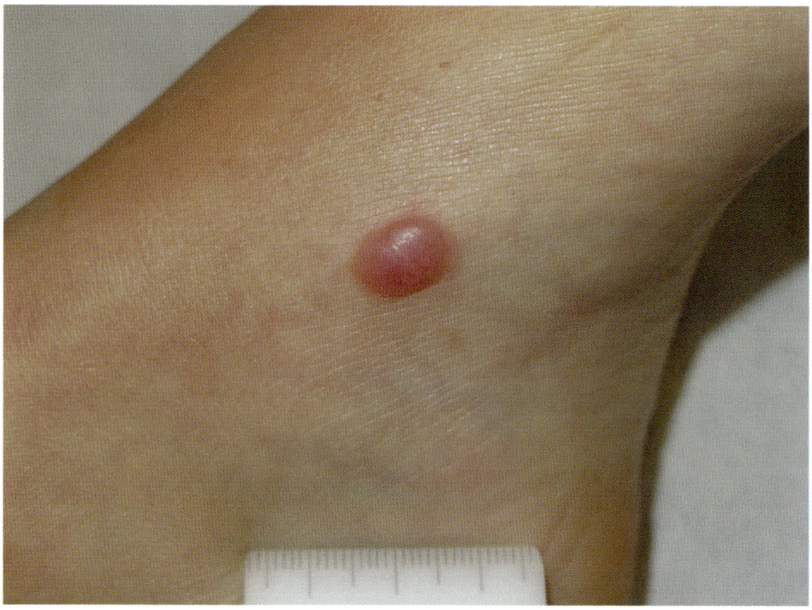

Fig. 5.37 Clinical close-up examination revealed a symmetric roundish nodule with a diameter of 1.2 cm without any sign of pigmentation

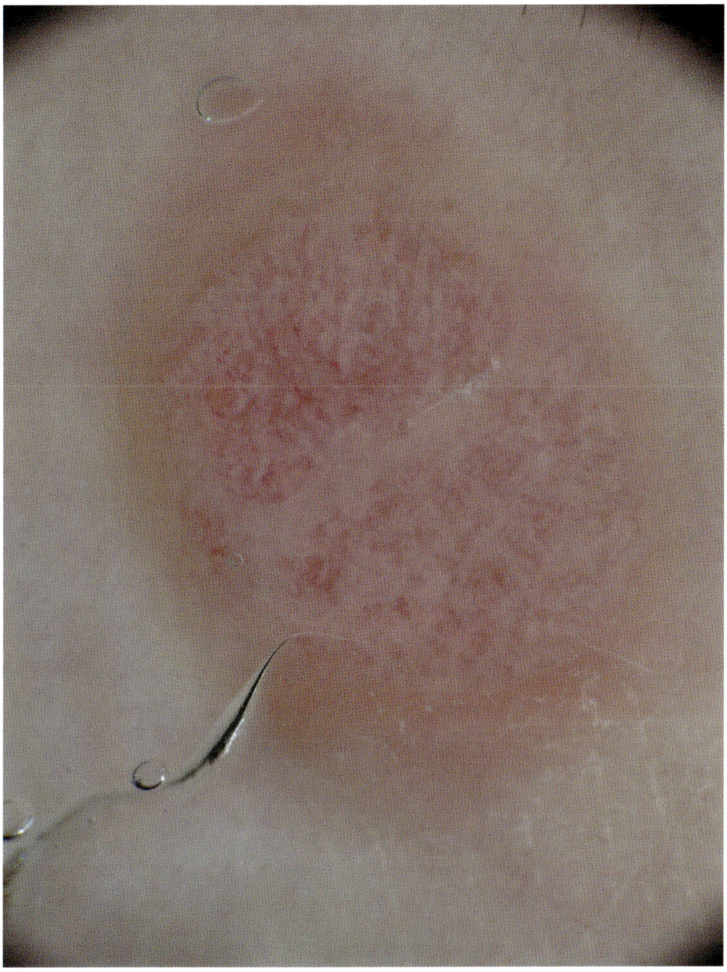

Fig. 5.38 Dermoscopy revealed multiple fine atypical vessels along with crystalline structures

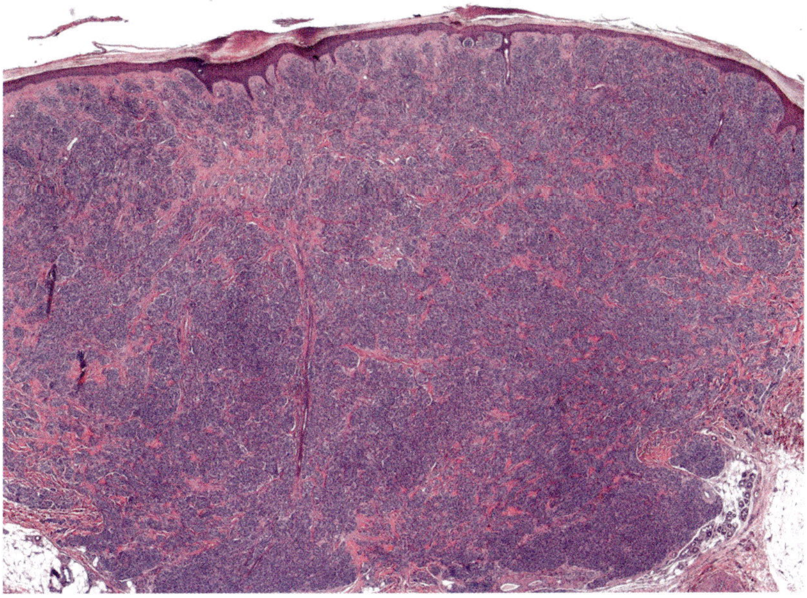

Fig. 5.39 At scanning, the lesion was a symmetrical dermo-hypodermal nodule

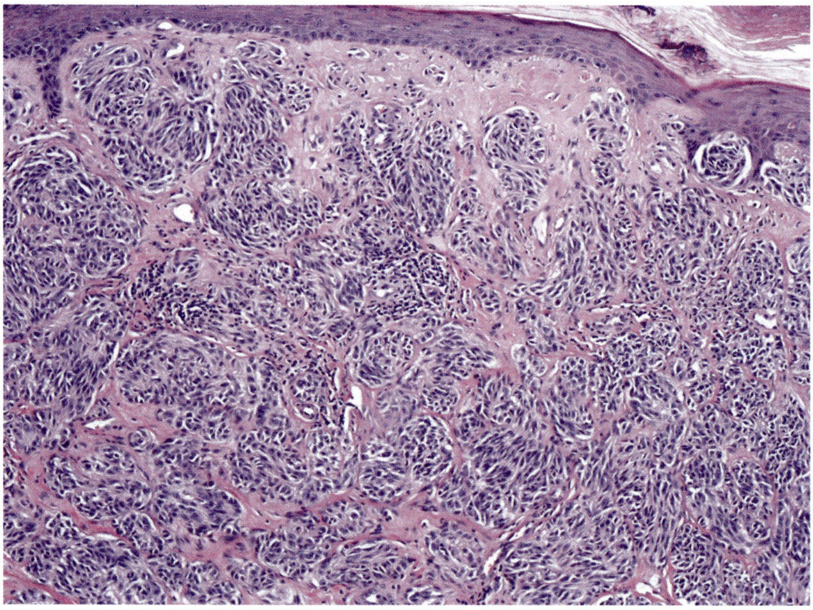

Fig. 5.40 It was composed of monomorphous spindle or epithelioid cells, arranged in a diffuse fashion without nesting. Some mitoses were noted in the deep dermis

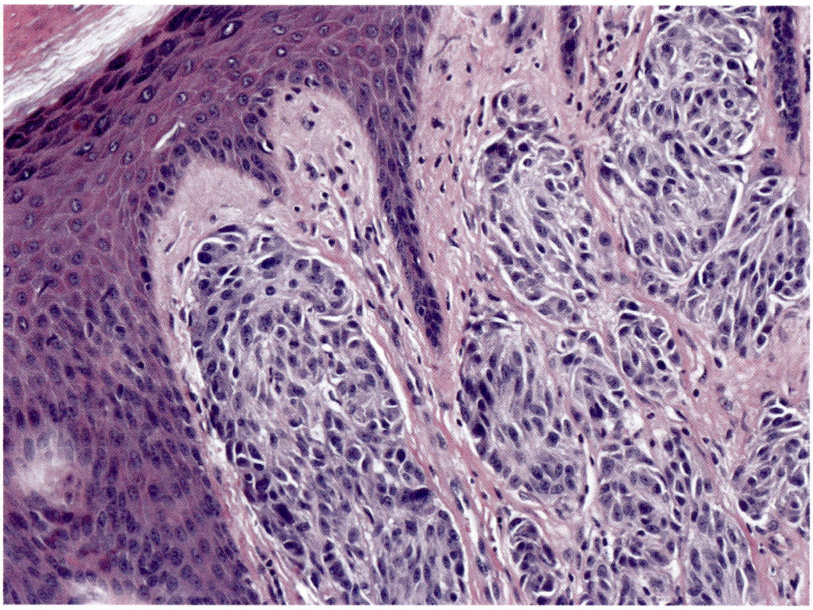

Fig. 5.41 The epithelioid cells grew in a diffuse pattern without infiltrating the epidermis

Case 9

Patient: Male, 36 years old

Anatomic site of the lesion: Left ankle

History of the lesion: Fast-growing lesion. The patient had no moles on the trunk and the extremities.

Management: Atypical vessels typified the dermoscopic pattern, and in one small area, a remnant of brown pigment was highly suggestive of a melanocytic lesion, possibly spitzoid. The lesion was excised to rule out melanoma.

Histopathologic diagnosis: Verrucous compound Spitz nevus

Key message: The diagnosis of red nodular lesions can be extremely challenging since under similar morphologic appearance, several distinct neoplasias have to be considered.

The location on the leg and the verrucous clinical aspect could have raised the suspicion of a verrucous Spitz nevus. However, the age of the patient (i.e., adult) and the fast-growing attitude along with the specific dermoscopic feature could not reliably exclude the diagnosis of amelanotic melanoma.

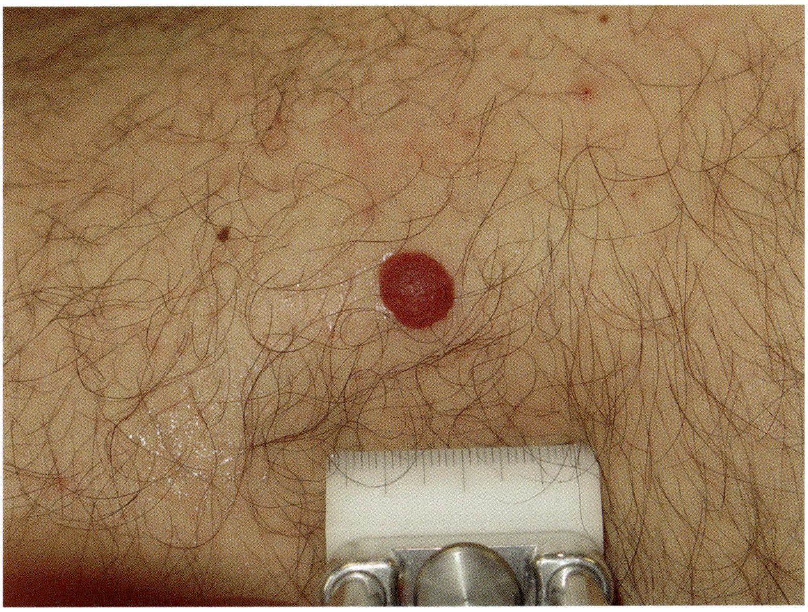

Fig. 5.42 Close-up clinical examination of the largest mole on his left ankle revealed a reddish papillomatous nodule

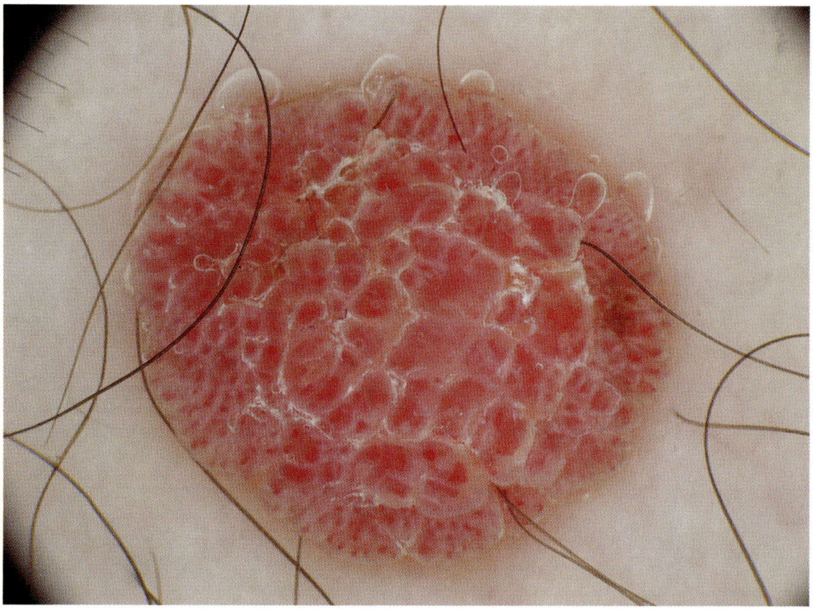

Fig. 5.43 Dermoscopically, the lesion was characterized by multiple and irregularly distributed dark-brown-to-black globules and multiple dotted vessels

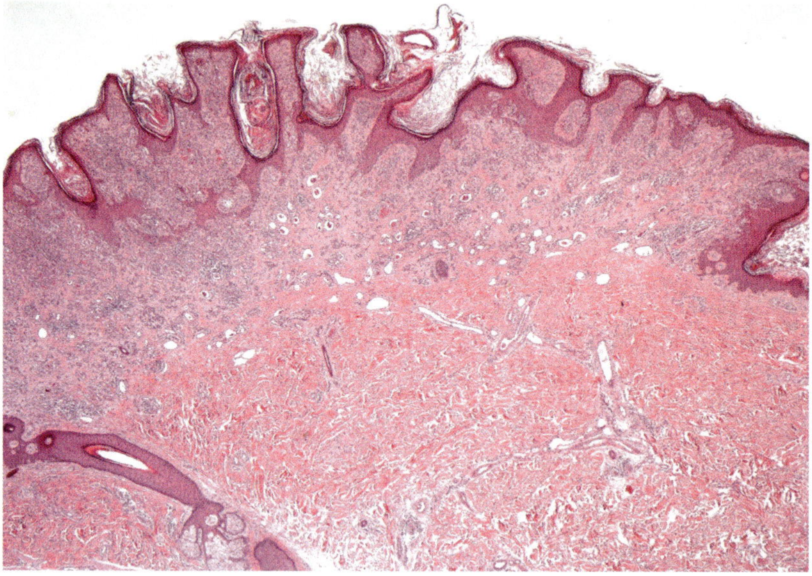

Fig. 5.44 The lesion had a striking verrucous silhouette with a horizontal deep border. The epidermis was hyperkeratotic

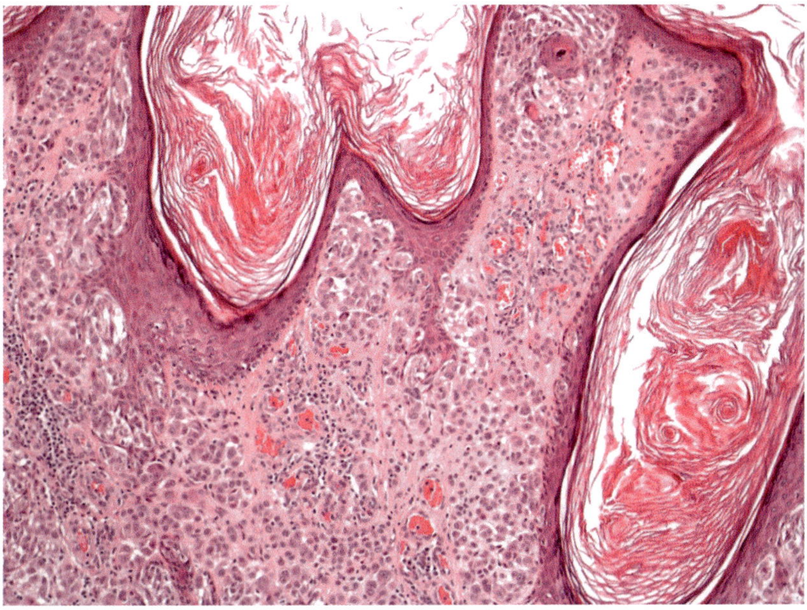

Fig. 5.45 The cells had a typical epithelioid appearance with abundant eosinophilic cytoplasm

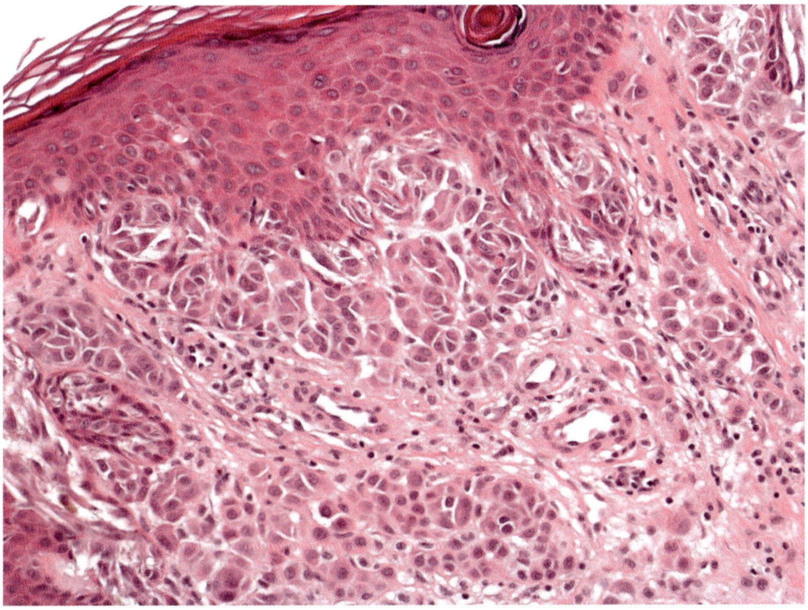

Fig. 5.46 The growth pattern was in discohesive irregular nests but the cells were rather monomorphic

Case 10

Patient: Male, 53 years old

Anatomic site of the lesion: Left thigh

History of the lesion: Newly appearing pigmented lesion

Management: Differential diagnoses included melanoma versus Spitz/Reed nevus.

Histopathologic diagnosis: Spitzoid melanoma. The lesion was sent to an expert consultant who rendered a diagnosis of nevoid melanoma.

Key message: Spitzoid-looking lesions should always be managed carefully since the distinction between Spitz nevus and melanoma is not feasible in some instances based on the dermoscopic aspect. In fact, those lesions might reveal overlapping dermoscopic features such as streaks and dark pigmentation as well as dotted vessels in both nevi and melanoma. A practical golden rule to not miss a melanoma is to excise any given spitzoid-looking lesions in adults after puberty. Reflectance confocal microscopy could be of help in some cases to improve the diagnostic accuracy for melanoma.

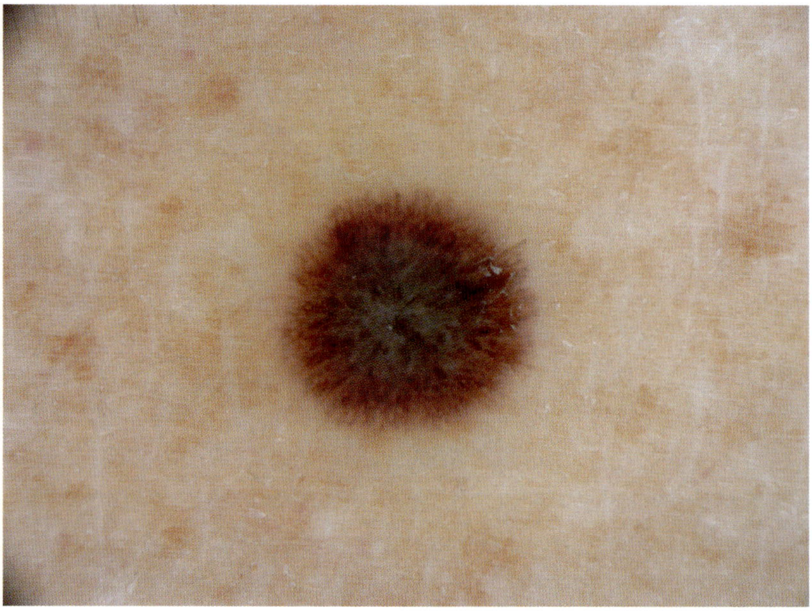

Fig. 5.47 Dermoscopy revealed a symmetric pigmented lesion with peripheral streaks and central blue-white veil

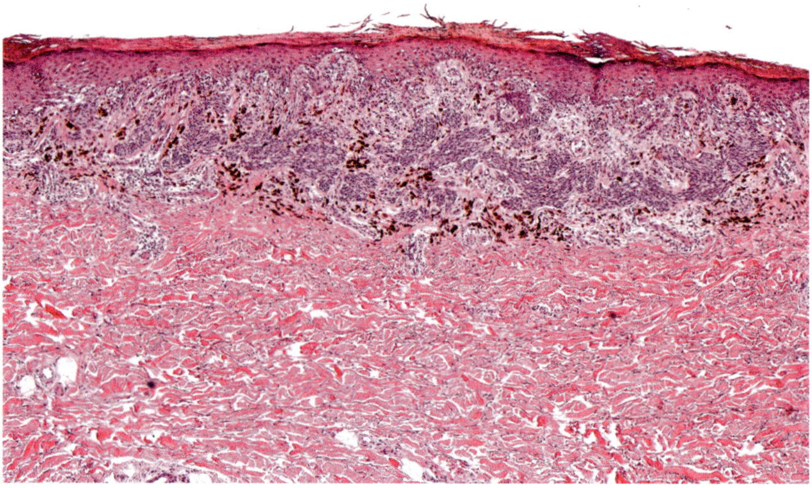

Fig. 5.48 Upon histology, the lesion was symmetric and relatively well defined, made up by spindle melanocytes. Many macrophages were present in the dermis

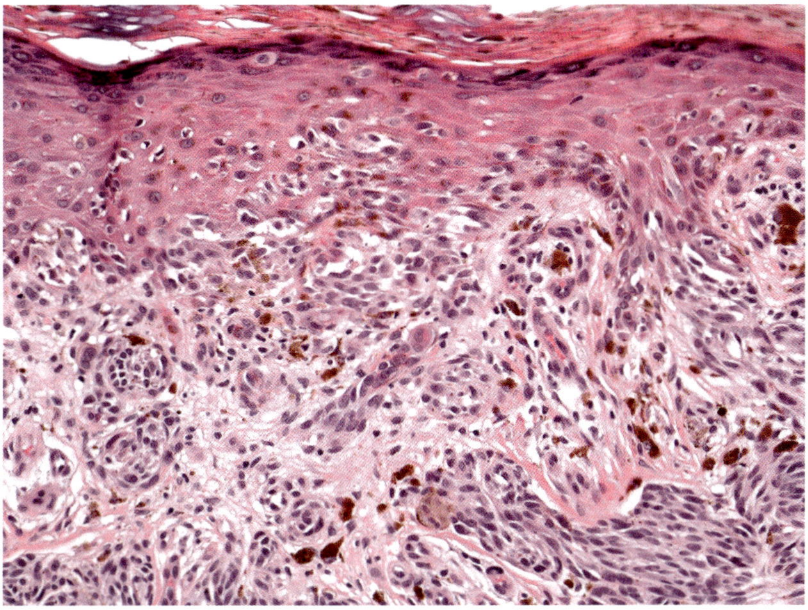

Fig. 5.49 Neoplastic cells were small and monomorphic. Pagetoid spread to the epidermis was remarkable

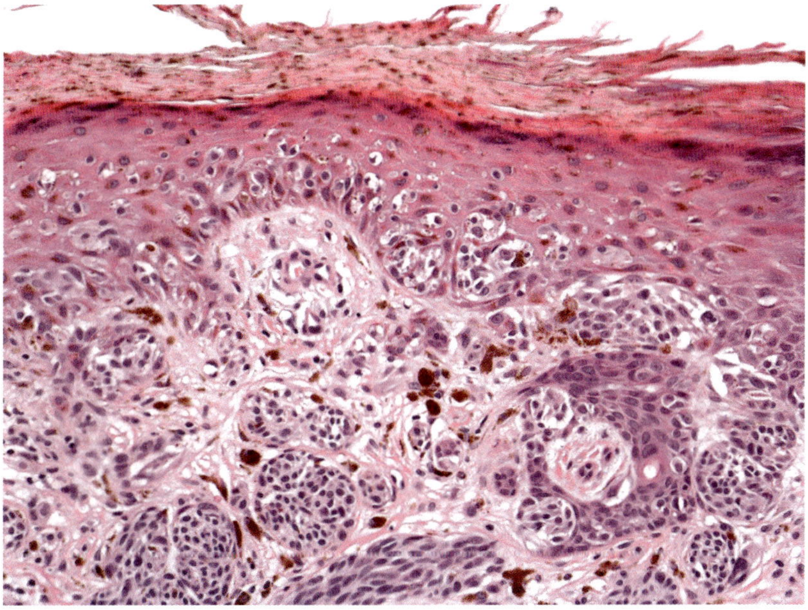

Fig. 5.50 In addition to the upward migration of melanocytes, many granules of pigment were seen within the stratum corneum

Suggested Reading

1. Argenziano G, Catricalà C, Ardigo M, et al. Seven-point checklist of dermoscopy revisited. Br J Dermatol. 2011;164:785–90.
2. Lallas A, Moscarella E, Longo C, et al. Likelihood of finding melanoma when removing a Spitzoid-looking lesion in patients aged 12 years or older. J Am Acad Dermatol. 2015;72:47–53.
3. Lallas A, Apalla Z, Ioannides D, et al. Update on dermoscopy of Spitz/Reed naevi and management guidelines by the International Dermoscopy Society. Br J Dermatol. 2017; doi:10.1111/bjd.15339.
4. Pedrosa AF, Lopes JM, Azevedo F, Mota A. **Spitz/** Reed nevi: a review of clinical-dermatoscopic and histological correlation. Dermatol Pract Concept. 2016;6(2):37–41.
5. Lott JP, Wititsuwannakul J, Lee JJ, Ariyan S, Narayan D, Kluger HH, Lazova R. Clinical characteristics associated with Spitz nevi and Spitzoid malignant melanomas: the Yale University Spitzoid Neoplasm Repository experience, 1991 to 2008. J Am Acad Dermatol. 2014;71(6):1077–82.
6. Semkova K, Lott JP, Lazova R. Clinicopathologic features and survival in Spitzoid malignant melanoma and conventional malignant melanoma. J Am Acad Dermatol. 2014;71(3):516–20.
7. Broganelli P, Titli S, Lallas A, Alaibac M, Annetta A, Battarra V, Brunetti B, Castagno I, Cavicchini S, Ferrari A, Ghigliotti G, Landi C, Manganoni A, Moscarella E, Pellacani G, Pizzichetta MA, Rosina P, Rubegni P, Satta R, Scalvenzi M, Stanganelli I, Stinco G, Zalaudek I, Zampieri P. Argenziano G. Spitz/Reed nevi: proposal of management recommendations by the Dermoscopy Study Group of the Italian Society of Dermatology (SIDeMaST). Dermoscopy Study Group of the Italian Society of Dermatology. G Ital Dermatol Venereol. 2014;149(5):601–6.
8. Moscarella E, Lallas A, Kyrgidis A, Ferrara G, Longo C, Scalvenzi M, Staibano S, Carrera C, Díaz MA, Broganelli P, Tomasini C, Cavicchini S, Gianotti R, Puig S, Malvehy J, Zaballos P, Pellacani G, Argenziano G. Clinical and dermoscopic features of atypical Spitz tumors: a multicenter, retrospective, case-control study. J Am Acad Dermatol. 2015;73(5):777–84.
9. Lallas A, Kyrgidis A, Ferrara G, Kittler H, Apalla Z, Castagnetti F, Longo C, Moscarella E, Piana S, Zalaudek I, Argenziano G. Atypical Spitz tumours and sentinel lymph node biopsy: a systematic review. Lancet Oncol. 2014;15(4):e178–83.
10. Gerami P, Busam K, Cochran A, Cook MG, Duncan LM, Elder DE, Fullen DR, Guitart J, LeBoit PE, Mihm MC Jr, Prieto VG, Rabkin MS, Scolyer RA, Xu X, Yun SJ, Obregon R, Yazdan P, Cooper C, Weitner BB, Rademaker A, Barnhill RL. Histomorphologic assessment and interobserver diagnostic reproducibility of atypical spitzoid melanocytic neoplasms with long-term follow-up. Am J Surg Pathol. 2014;38(7):934–40.
11. Moreno-Ramirez D, Rios-Martin JJ, Ferrandiz L, Camacho FM. Dermoscopic appearance of dermatofibroma-like Spitz nevus. J Am Acad Dermatol. 2009;61(5):910–2.
12. Ferrara G, Gianotti R, Cavicchini S, Salviato T, Zalaudek I, Argenziano G. Spitz nevus, Spitz tumor, and spitzoid melanoma: a comprehensive clinicopathologic overview. Dermatol Clin. 2013;31(4):589–9.
13. Ferrara G, Argenziano G, Soyer HP, Chimenti S, Di Blasi A, Pellacani G, Peris K, Piccolo D, Rubegni P, Seidenari S, Staibano S, Zalaudek I, De Rosa G. The spectrum of Spitz nevi: a clinicopathologic study of 83 cases. Arch Dermatol. 2005;141(11):1381–7.
14. Moscarella E, Al Jalbout S, Piana S, Argenziano G, Lallas A, Longo C, Hofmann-Wellenhof R, Zalaudek I. The stars within the melanocytic garden: unusual variants of Spitz naevi. Br J Dermatol. 2015;172(4):1045–51.
15. Guida S, Pellacani G, Cesinaro AM, Moscarella E, Argenziano G, Farnetani F, Bonamonte D, Longo C. Spitz naevi and melanomas with similar dermoscopic patterns: can confocal microscopy differentiate? Br J Dermatol. 2016;174(3):610–6.

6 Lesions with Regression

Regression of pigmented skin tumors represents a peculiar biologic phenomenon with not fully elucidated significance.

Regression is known to occur in a considerable percentage of melanomas. The question of whether the presence of regression affects melanoma prognosis remains controversial. In some studies, melanomas undergoing regression were associated with a better prognosis. The possible explanation was that regression is a result of a successful immunologic response of the host against the tumor. However, other studies were not able to find any prognostic benefit of regression, and, finally, some others reported that regression might predict even a poorer prognosis. Interestingly, cases of metastatic melanoma of unknown primary tumor do exist, and it has been proposed that at least part of them represent a result of complete regression of the primary melanoma. If this assumption is correct, then it definitely suggests that regression cannot be considered a positive prognostic predictor.

Both benign and malignant melanocytic and non-melanocytic tumors might undergo regression. Therefore, the presence of structures suggestive of regression is not necessarily predictive of a malignant tumor.

Two main dermoscopic patterns of regression do exist: white scar-like depigmentation, histopathologically corresponding to fibrosis, and blue-gray granules (peppering), histopathologically corresponding to melanophages. A single lesion might display one or both of the regression patterns.

Regression structures are included in the list of melanoma-specific criteria, and, therefore, their detection should be considered as suggestive of melanoma. However, it is true that also nevi often undergo an involution process, which, from a dermoscopic point of view, is very similar to regression.

The discrimination between regression (of melanoma) and involution (of nevi) has been suggested to be based on the intensity of the white color. Specifically, regression is described as white scar-like areas, meaning that the color of the regressive part of the lesion is whiter than the healthy skin surrounding the lesion. In contrast, involution is seen as depigmentation, with the involuting areas acquiring gradually the color of the healthy skin, but not whiter. However, the assessment of the intensity of white color is not always easy and is rather subjective. Moreover, melanophages (dermoscopically seen as blue-tray granules) are present in both processes (regression and involution). Thus, the discrimination between regression and involution is often problematic.

As a rule, nevi are morphologically symmetric even when they undergo involution/regression. Most often a nevus involutes either from the center to the periphery or the opposite, while melanoma typically undergoes an asymmetric regression. Moreover, nevi usually display only one type of regression (white or blue), while melanoma often exhibits both.

C. Longo et al., *Atlas of Diagnostically Challenging Melanocytic Neoplasms*,
https://doi.org/10.1007/978-3-319-48653-6_6

Seborrheic keratoses and solar lentigines also undergo regression (lichen planus-like keratosis—LPLK), which is dermoscopically very similar to the one seen in melanocytic tumors. The only difference might be the diameter of the blue-gray structures, which often are not granules but dots/globules. If regression affects only part of the lesion, the remaining part may display dermoscopic criteria allowing the recognition of the tumor. When regression is extensive, LPLK cannot be safely discriminated from fully regressed melanoma.

It has been suggested that the management of regressive lesions should be based on the extent of the regression within the lesion, with the probability of melanoma increasing with the extent of the regression. The so-called tumoral melanosis, histologically represented by sheets and nodules of hyperpigmented melanophages, has been strongly associated with melanoma regression more than with nevi involution. As the presence of extensive regression is a highly confounding feature because it does not allow any safe conclusion on what the lesion initially was, the only safe management of such lesions would be excision.

Case 1

Patient: Male, 48 years old

Anatomic site of the lesion: Chest

History of the lesion: Long-standing lesion

Management: The lesion was excised to rule out a melanoma.

First histopathologic diagnosis: Junctional nevus

After clinicopathologic correlation, the lesion was sent to consultation to three expert dermatopathologists: Their diagnoses were in situ melanoma and junctional nevus. A definite diagnosis was not reached.

Key message: In thin atypical lesion, some histopathological controversies still exist. A correct pragmatical approach is excision with safe margins. The overall dermoscopic aspect with asymmetric distribution of colors and the presence of regression covering more than 50% of the lesion is highly suspicious, and thus surgical excision is mandatory.

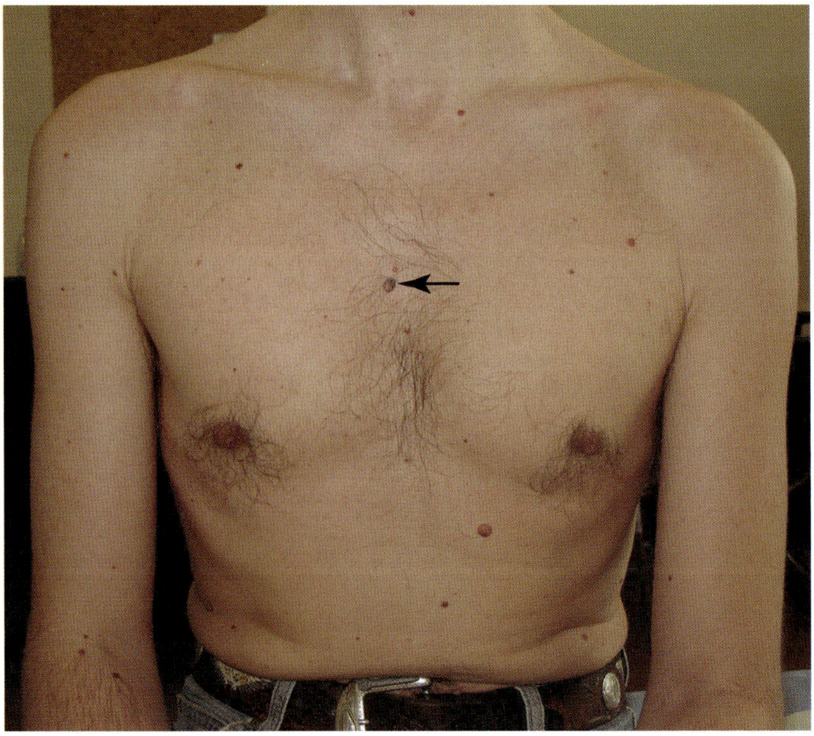

Fig. 6.1 Large lesion in the context of a patient with moderate number of nevi

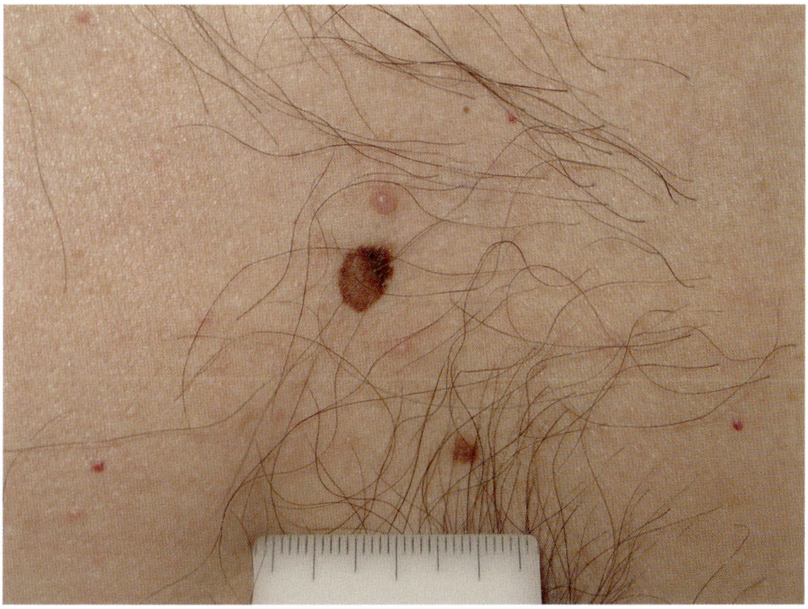

Fig. 6.2 Flat pigmented lesion with a diameter of more than 6 mm

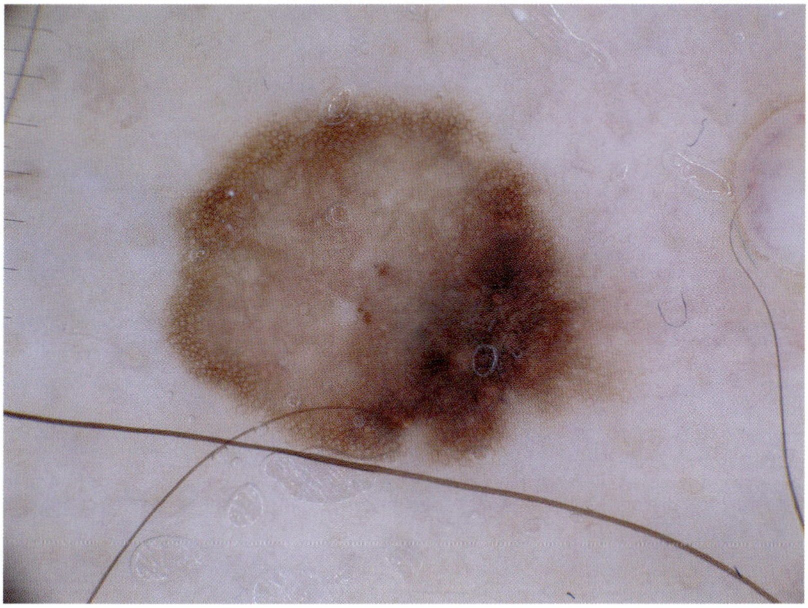

Fig. 6.3 Dermoscopically the lesion showed a symmetric atypical pigmented network with whitish regression

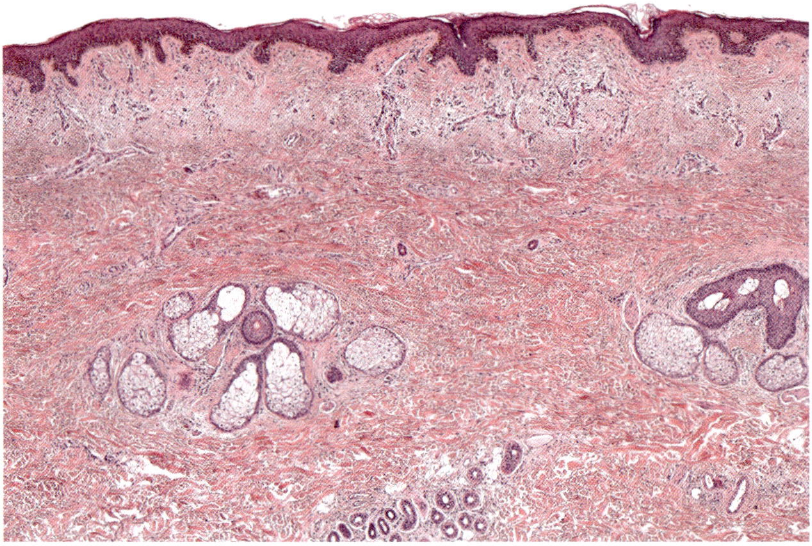

Fig. 6.4 On histology, the lesion consisted of a continuous melanocytic intraepidermal proliferation, mainly located in the basal layers of the epidermis

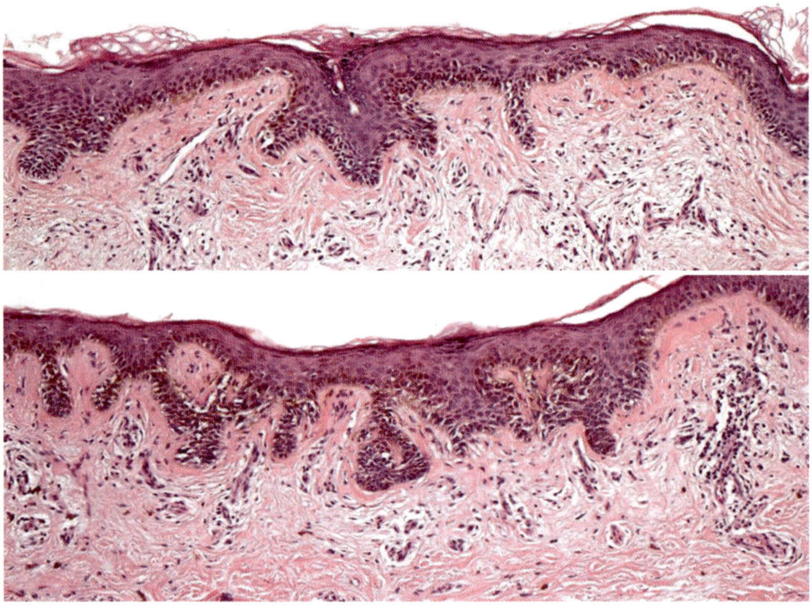

Fig. 6.5 Cytological atypia was minimal. The dermis was fibrotic; many vessels and rare inflammatory cells were present

Case 2

Patient: Male, 67 years old

Anatomic site of the lesion: Upper back

History of the lesion: Unknown

Management: The dermoscopic findings were interpreted as suggestive of a lesion with regression, and, therefore, the lesion was excised to rule out melanoma.

Histopathologic diagnosis: Tumoral melanosis, most likely due to completely regressed melanoma.

The patient subsequently underwent a surgical excision of an axillary lymph node, with a pathologic report of lymph node metastasis of melanoma.

Key message: Lesions undergoing complete regression are extremely challenging because there are no remnants of melanocytic or non-melanocytic proliferations, and the histologic diagnosis is basically based on the arrangement of fibrosis and melanophages as in this peculiar case.

From a clinical point of view, the present case displayed only blue colors in the absence of any other dermoscopic diagnostic structures. To further complicate the matter, a tiny remnant of seborrheic keratosis was noted, leading toward possible diagnosis of benign lesion in regression. However, the presence of an asymmetric and peripherally located bluish veil was considered suspicious, and thus the lesion was excised. Since histopathology did not report a clear-cut diagnosis, follow up was recommended and a narrow margin excision was performed. Lately, a lymph node metastasis was detected making possible a diagnosis of melanoma a posteriori.

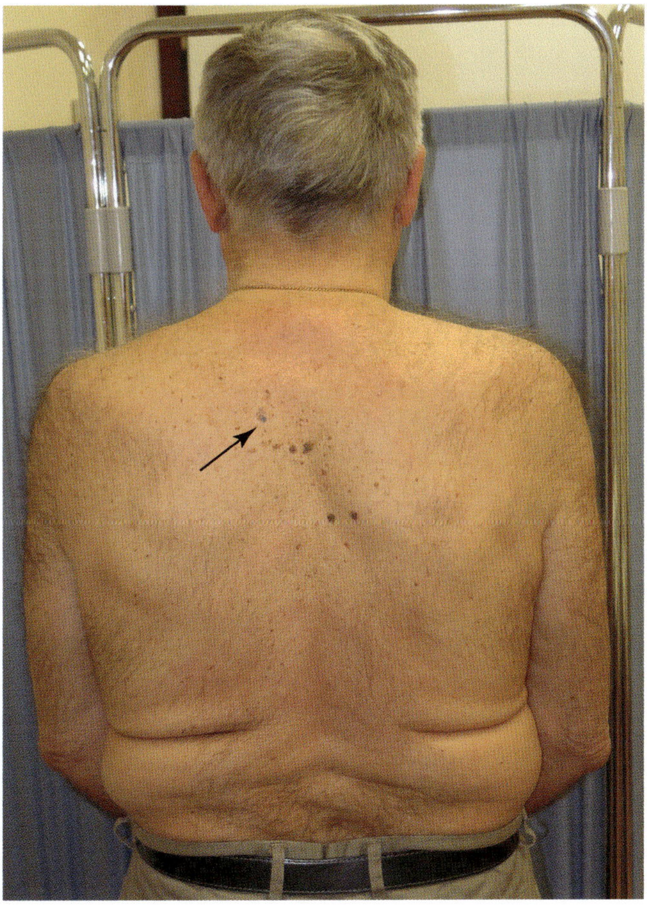

Fig. 6.6 Upper back of a patient with several seborrheic keratoses

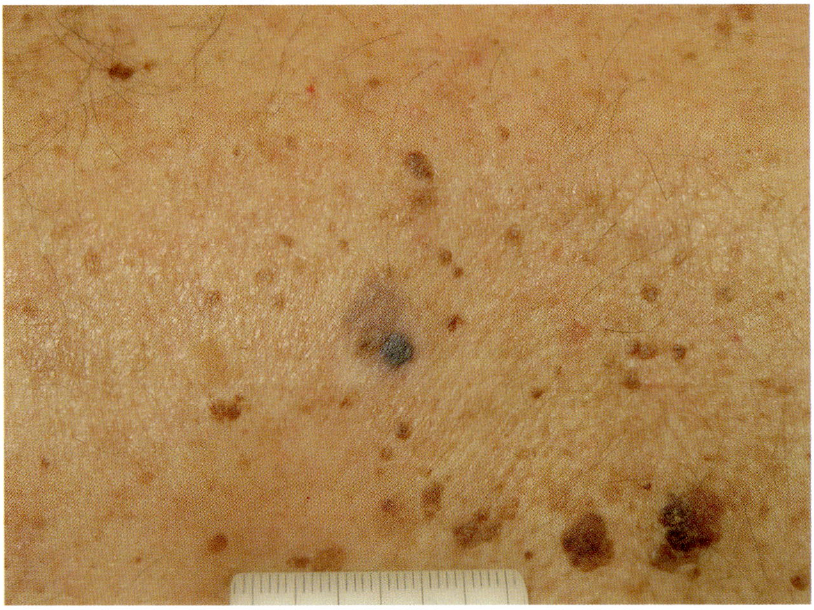

Fig. 6.7 A flat asymmetric pigmented lesion with a darker eccentric area was present within a sun damaged skin

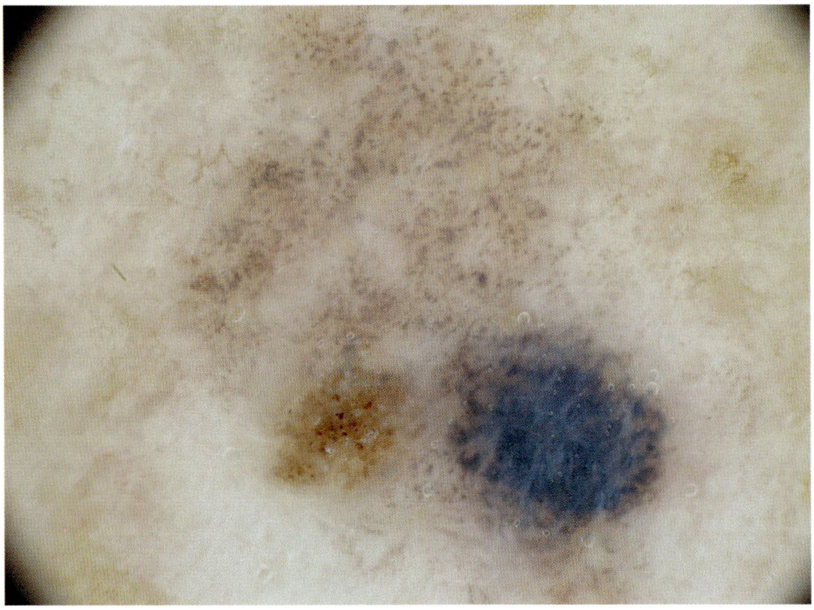

Fig. 6.8 Dermoscopically, the lesion displayed regression with bluish granules and in the lower portion a blue veil. Notably, in proximity of the blue veil, there was an area referable to seborrheic keratosis

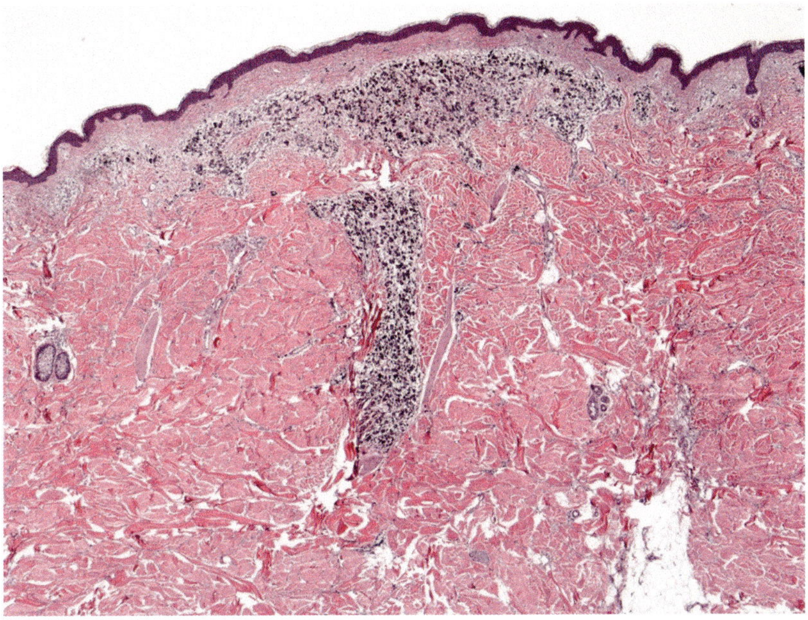

Fig. 6.9 On histology, the lesion was a dumbbell aggregate of heavily pigmented melanophages. The overlying epidermis was unremarkable. No residual melanocytic tumor could be seen

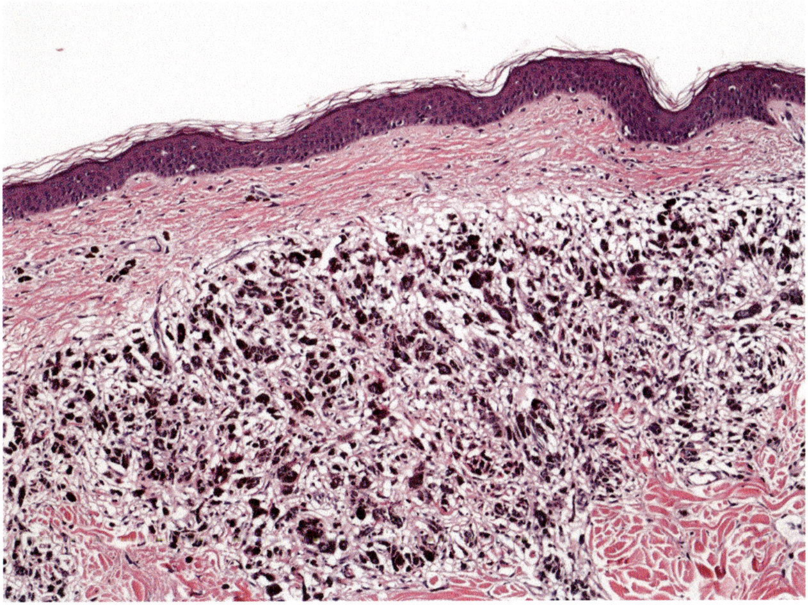

Fig. 6.10 Abundant melanin-containing macrophages were intermixed with lymphocytes and plasma cells

Case 3

Patient: Female, 44 years old

Anatomic site of the lesion: Left lower back

History of the lesion: Unknown

Management: Because of the asymmetry of colors and structures, the lesion was excised to rule out melanoma.

Histopathologic diagnosis: Invasive melanoma arising on congenital-type compound nevus

Key message: Regression can be present in several melanocytic and non-melanocytic lesions. In melanocytic proliferations, regression shows a peculiar arrangement. In fact in nevi, it is usually located in the center of the lesion that is rather symmetric, and no other clues are seen. Conversely, in melanoma, regression could be asymmetrically distributed as in this case and associated to atypical network. However, in cases showing regression in over 50% of the lesion, excision remains the golden rule.

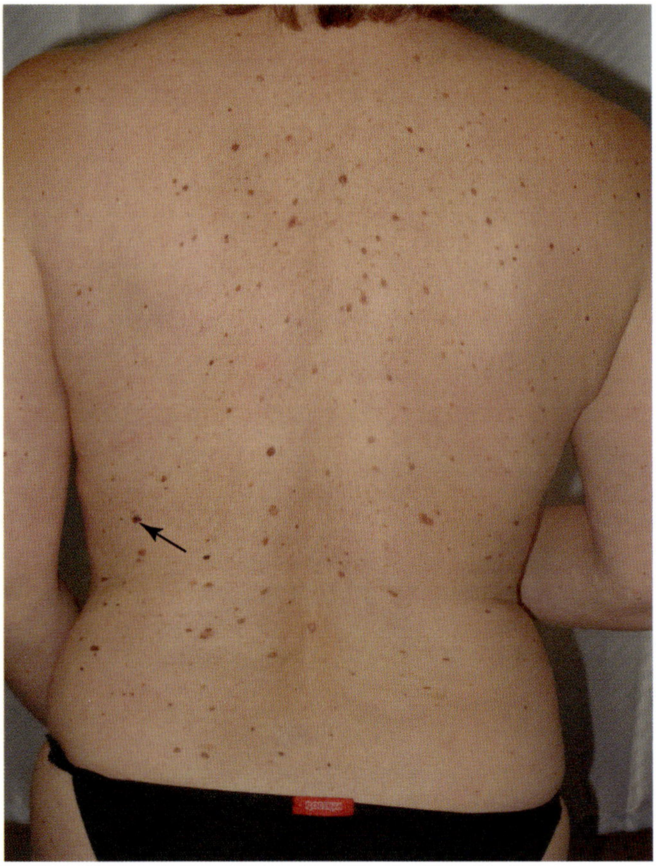

Fig. 6.11 On the back multiple moles could be seen

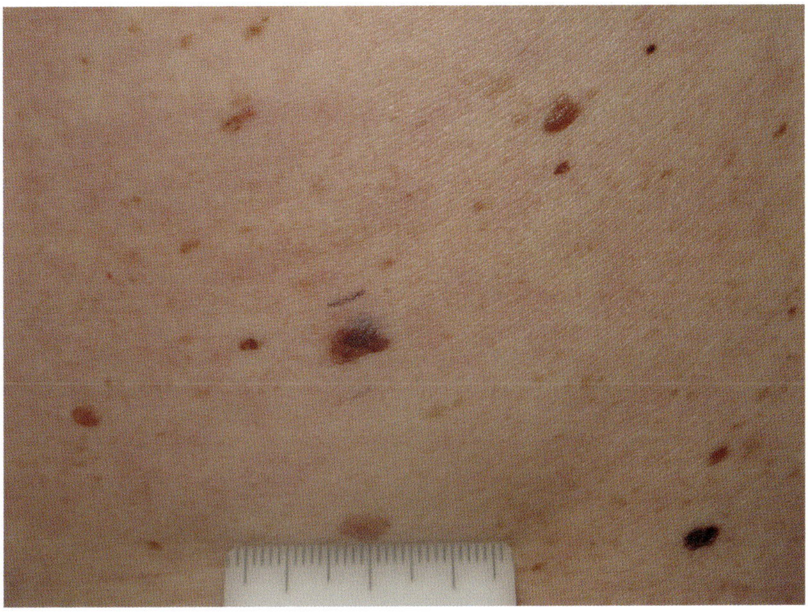

Fig. 6.12 A flat pigmented macule of dark-brown color with irregular borders and a diameter of 7 mm. At the upper right part of the lesion, a bluish color was present, resulting in an overall asymmetry for color distribution

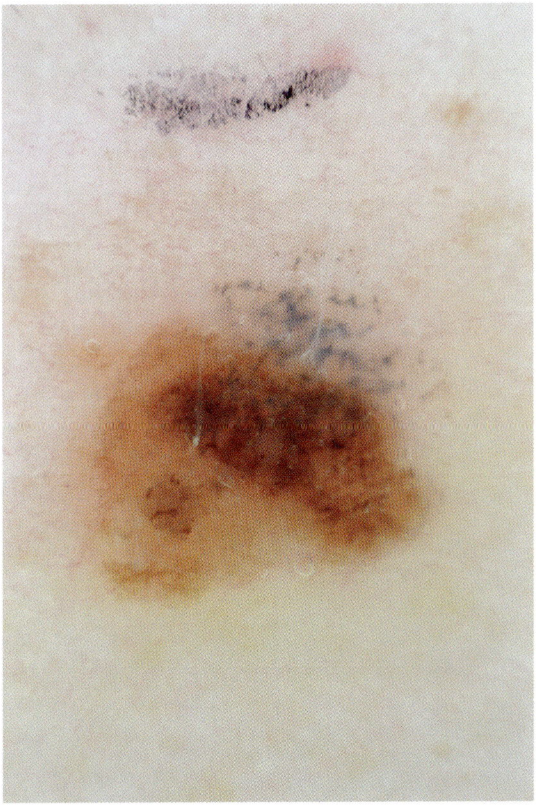

Fig. 6.13 Dermoscopy revealed a globular central portion with peripheral network. On the right side, regression of the network was present

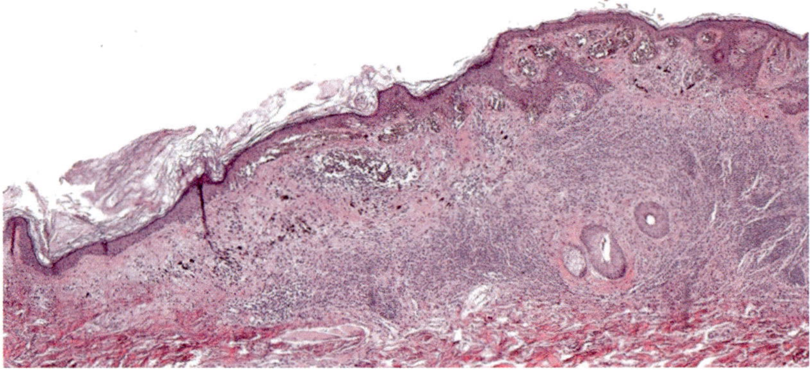

Fig. 6.14 At low-power view, the lesion was asymmetrical, but it appeared mostly as a congenital-type, hypopigmented nevus. Peripherally, at the junction and in the superficial dermis, however, some confluent nests of atypical pigmented melanocytes were easily appreciable

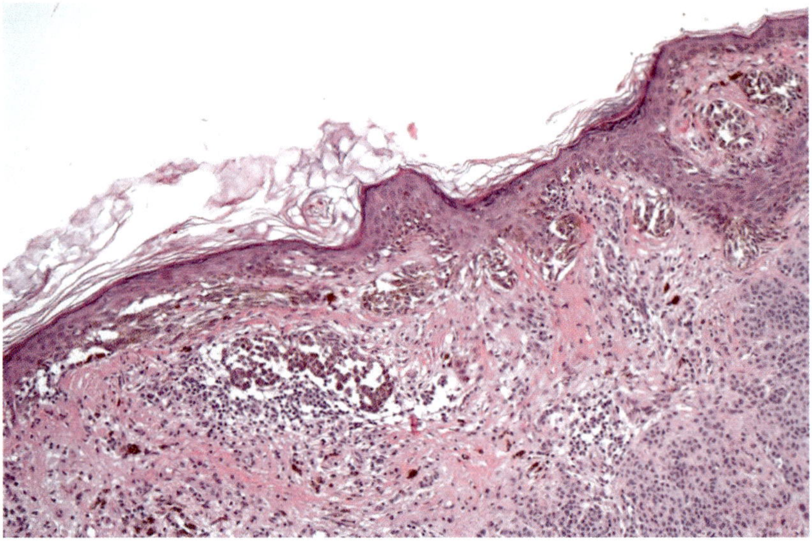

Fig. 6.15 Atypical melanocytes were strikingly different from those belonging to congenital nevus (right lower corner)

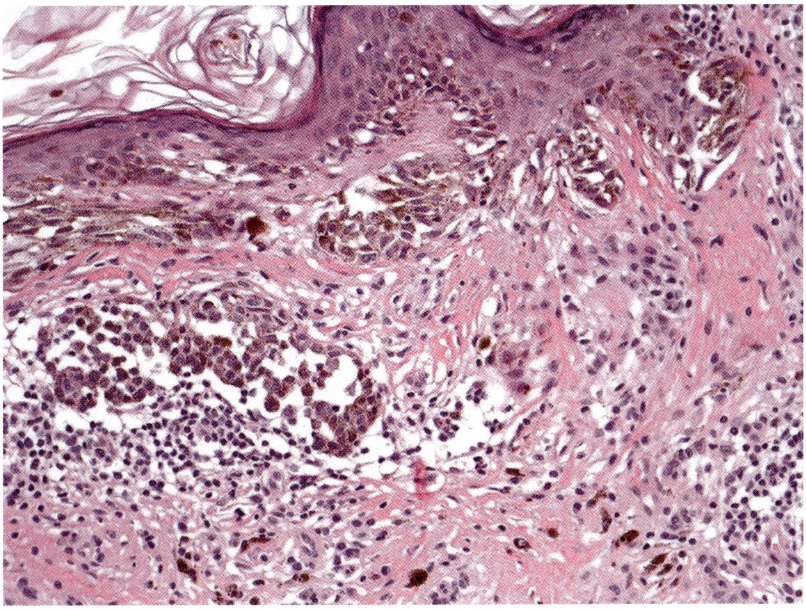

Fig. 6.16 Atypical melanocytes invaded the superficial dermis in a fibrotic and inflammatory background

Suggested Reading

1. Zalaudek I, Lallas A, Longo C, Moscarella E, Tiodorovic-Zivkovic D, Ricci C, Albertini G, Argenziano G. Problematic lesions in the elderly. Dermatol Clin. 2013;31(4):549–64.
2. Ribero S, Moscarella E, Ferrara G, Piana S, Argenziano G, Longo C. Regression in cutaneous melanoma: a comprehensive review from diagnosis to prognosis. J Eur Acad Dermatol Venereol. 2016;30(12):2030–7.
3. Shoo BA, Sagebiel RW, Kashani-Sabet M. Discordance in the histopathologic diagnosis of melanoma at a melanoma referral center. J Am Acad Dermatol. 2010;62(5):751–6.
4. Lallas A, Apalla Z, Moscarella E, Zalaudek I, Tzellos T, Lefaki I, Cota C, Argenziano G. Extensive regression in pigmented skin lesions: a dangerous confounding feature. Dermatol Pract Concept. 2012;2(2):202a08.
5. Ng SH, Chave TA. Tumoral melanosis as a manifestation of a completely regressed primary melanoma with metastases. Br J Dermatol. 2006;155(3):627–8.
6. Tannous ZS, Mihm MC Jr, Sober AJ, Duncan LM. Congenital melanocytic nevi: clinical and histopathologic features, risk of melanoma, and clinical management. J Am Acad Dermatol. 2005;52(2):197–203.
7. Ribero S, Moscarella E, Ferrara G, Piana S, Argenziano G, Longo C. Regression in cutaneous melanoma: a comprehensive review from diagnosis to prognosis. J Eur Acad Dermatol Venereol. 2016;30(12):2030–7.
8. Buçard AM, Marques-da-Costa J, Argenziano G, Moscarella E, Specchio F, Longo C. Regressive scalp lesions: dermoscopic and confocal clues. J Am Acad Dermatol. 2015;72(1 Suppl):S27–9.
9. Zalaudek I, Argenziano G, Ferrara G, Soyer HP, Corona R, Sera F, Cerroni L, Carbone A, Chiominto A, Cicale L, De Rosa G, Ferrari A, Hofmann-Wellenhof R, Malvehy J, Peris K, Pizzichetta MA, Puig S, Scalvenzi M, Staibano S, Ruocco V. Clinically equivocal melanocytic skin lesions with features of regression: a dermoscopic-pathological study. Br J Dermatol. 2004;150(1):64–71.

7 Acral Lesions

Although rare, in nonwhite populations, acral melanoma is the most prevalent type of melanoma in Asia, Latin America, Africa, and the Middle East. In Japan, acral melanoma accounts for approximately 50% of all melanomas, and about 10% of the population in Japan presents with acral nevus.

While acral lentiginous melanoma is easy to diagnose clinically in advanced stages, as it usually reaches several centimeters in axis, the differentiation of an early acral melanoma from a benign melanocytic nevus can be extremely challenging because both are observed as a brownish-black macule.

Dermoscopy is an important tool in the evaluation of acral lesions. The most common dermoscopic pattern in acral nevus is the parallel furrow pattern and its variants, while the dermoscopic feature most frequently detected in early acral melanoma is the parallel ridge pattern.

However, it has recently been shown that approximately one-third of acral melanomas do not display a parallel ridge pattern, thus rendering their accurate diagnosis more difficult. Some additional features, such as irregular diffuse pigmentation and other nonsite-specific melanoma criteria, have also been reported to characterize acral melanoma. In general, the presence of asymmetry of colors and asymmetry of structures and the presence of parallel ridge pattern are good predictors of melanoma diagnosis.

The histologic diagnosis of early acral lentiginous melanoma is often difficult, and it always requires a clinicopathologic correlation with age of the patient and size and evolution of the lesions. The first phase is the proliferation of single cells along the basal epidermal layer; these neoplastic melanocytes, with time, form irregular and non-cohesive nests, an important clue for the differential diagnosis with nevi, which usually show well-circumscribed, regularly distributed nests.

As it is recognized that a melanoma in situ can be exceedingly similar to a banal nevus, the complete excision of any clinically suspect lesion, especially in adults, is mandatory. In this chapter, three cases of acral melanocytic lesions arising in adulthood are presented to highlight the different possible scenarios in these special sites.

C. Longo et al., *Atlas of Diagnostically Challenging Melanocytic Neoplasms*,
https://doi.org/10.1007/978-3-319-48653-6_7

Case 1

Patient: Male, 46 years old

Anatomic site of the lesion: Second finger, left hand

History of the lesion: Since several years. Previously the lesion was diagnosed as "nevus" congenital type.

Management: The clinical and dermoscopic images were highly suggestive for melanoma. A punch biopsy was therefore obtained.

Histopathologic diagnosis: Invasive melanoma in acral skin

Key message: The overall clinical and dermoscopic pattern of this lesion along with its peculiar location on the transition area of the acral site is clear-cut diagnostic for melanoma. However, the lesion was neglected both by the patient and the doctors. Intuitively this could be related to the relatively young age of the patient when the lesion appeared and to its slow growing attitude. In order not to miss a melanoma, clinicians should be aware that large-sized asymmetric lesions should be always considered suspiciously and thus biopsied.

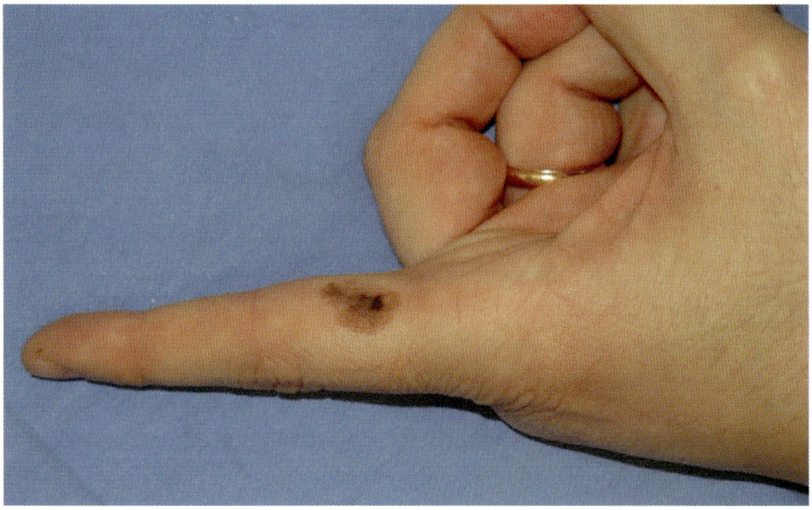

Fig. 7.1 The clinical examination revealed a solitary large pigmented macule with irregular borders. The color ranged from light to dark brown. Clinically the lesion was ABCD positive

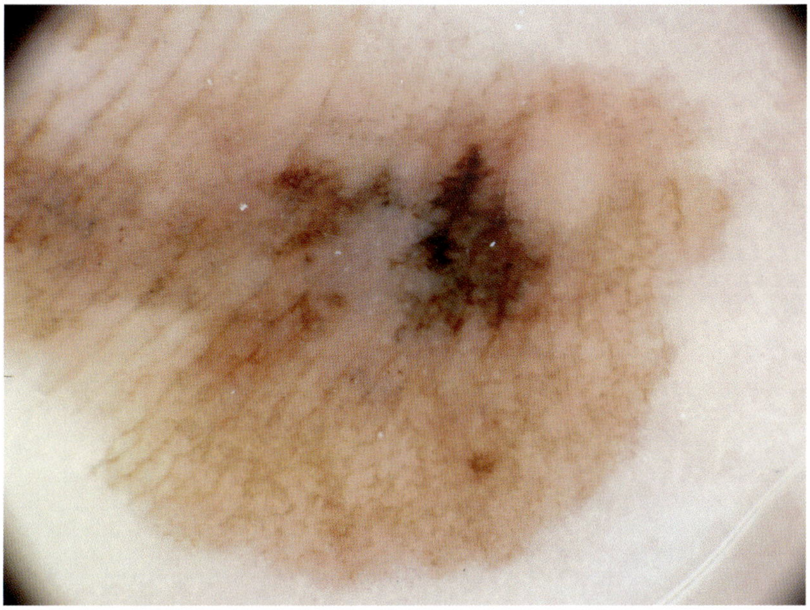

Fig. 7.2 Dermoscopy revealed an uneven distribution of colors and structures. A parallel pattern was seen with a darker area in which acrosyringia were completely obscurated

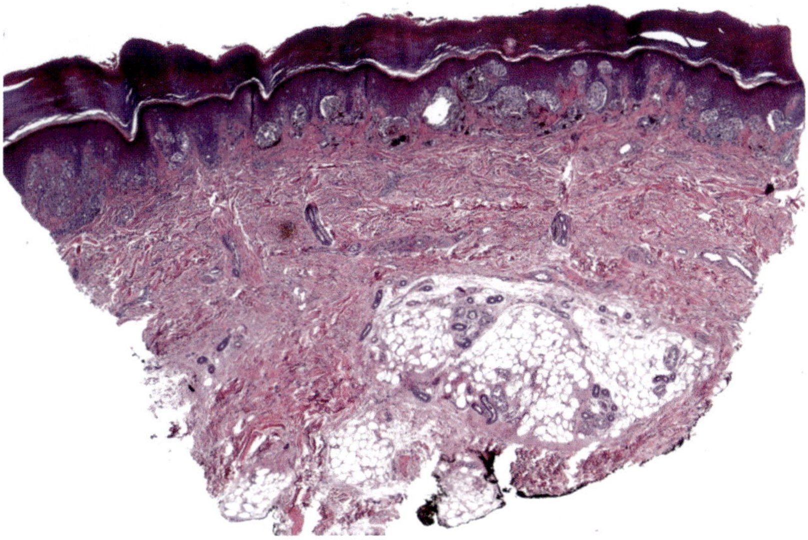

Fig. 7.3 At low-power view, there was an irregular, nested proliferation of clear melanocytes along the dermoepidermal junction. The epidermal was hyperkeratotic, as expected in an acral site

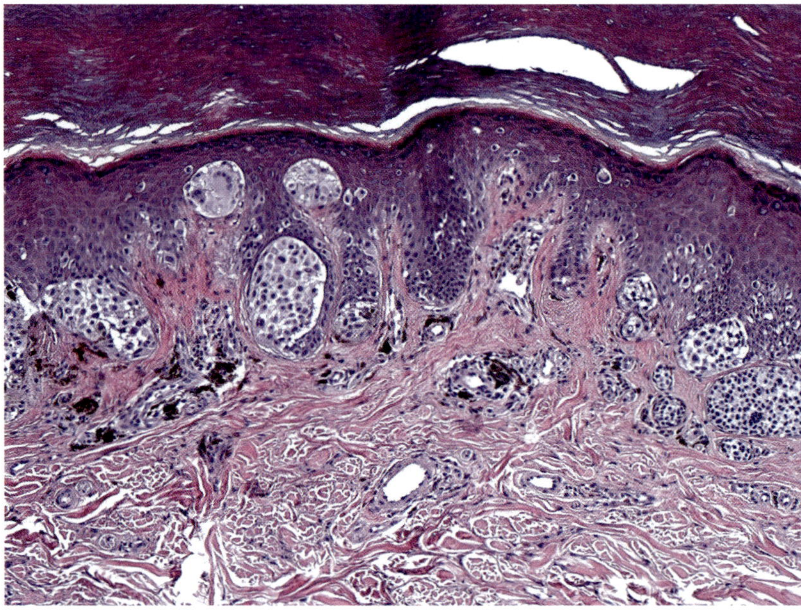

Fig. 7.4 Melanocytes were markedly atypical. They infiltrated the epidermis in nests and in single units

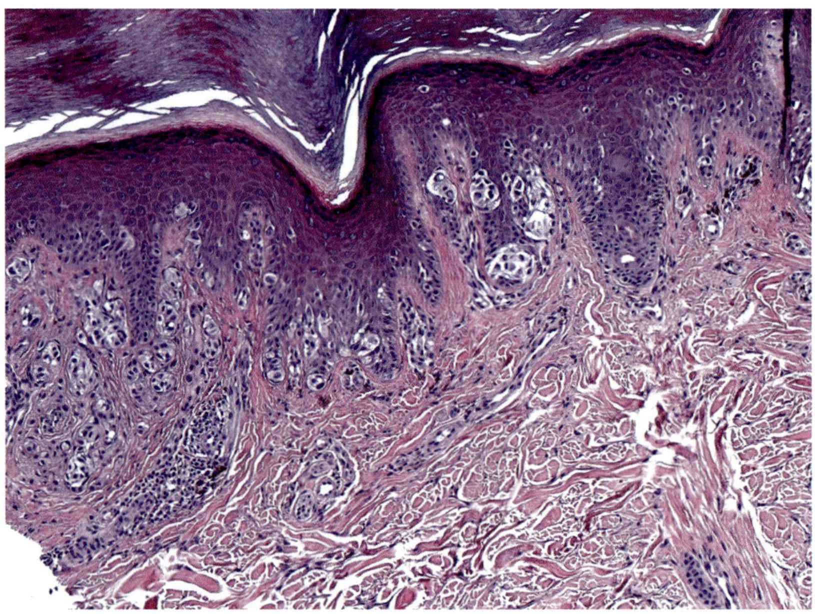

Fig. 7.5 On the left, an invasive dermal focus of melanoma could be appreciated

Case 2

Patient: Male, 70 years old

Anatomic site of the lesion: Acral, left foot

History of the lesion: Unknown

Management: The clinical and dermoscopic images were highly suggestive for melanoma.

Histopathologic diagnosis: Early in situ melanoma in acral skin

Key message: The distinction between acral melanomas and nevi in palms and soles can be particularly challenging, especially in early stages and in punch biopsy. As a general rule, a lesion appearing and enlarging in the elderly, even if histologically extraordinary bland, must be regarded as a melanoma. From a dermoscopic point of view, the presence of fibrillar pattern should be regarded with suspicion, and at least a digital monitoring should be performed.

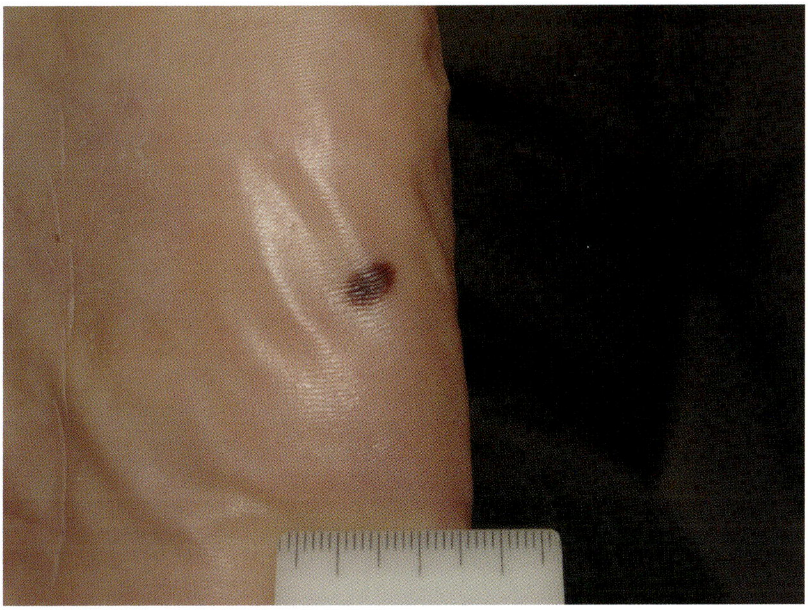

Fig. 7.6 Clinical examination revealed a solitary pigmented macule

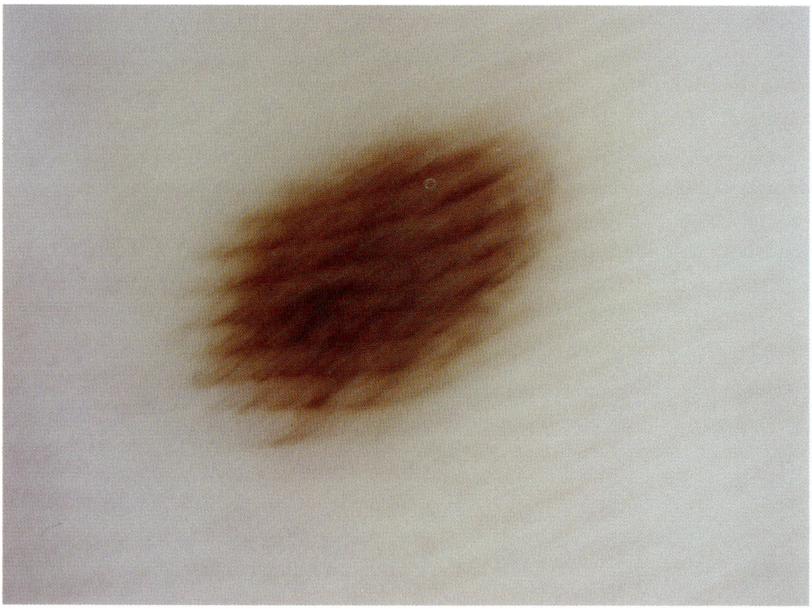

Fig. 7.7 Dermoscopy revealed a fibrillar pattern with color ranging from light to dark brown

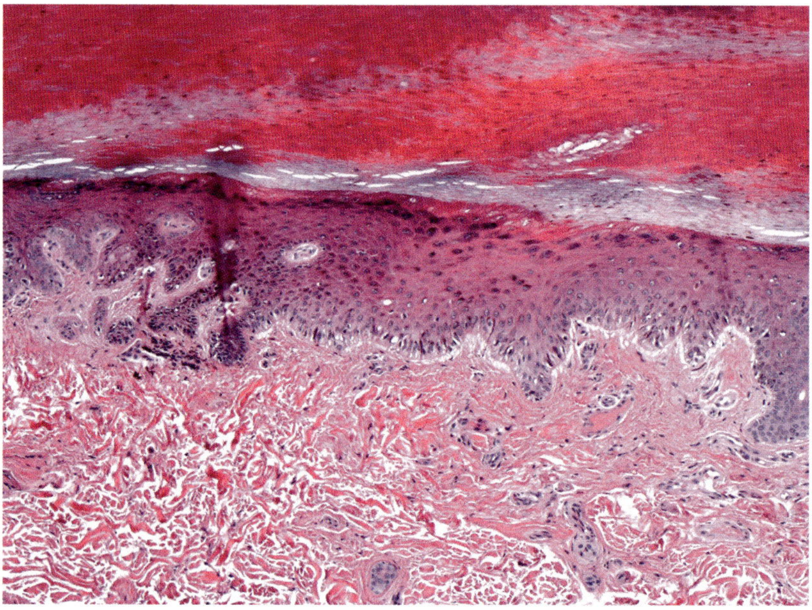

Fig. 7.8 On histology, major changes involved the epidermis that showed irregular hyperplasia and marked acanthosis

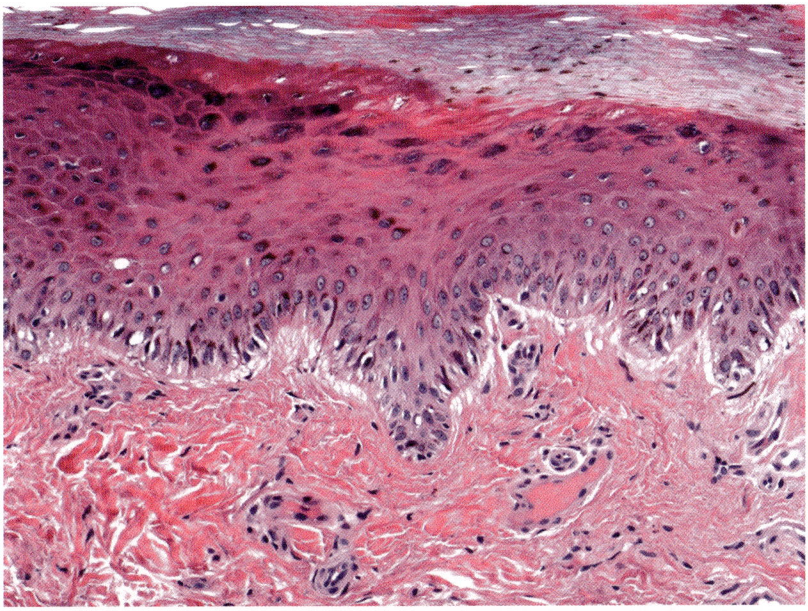

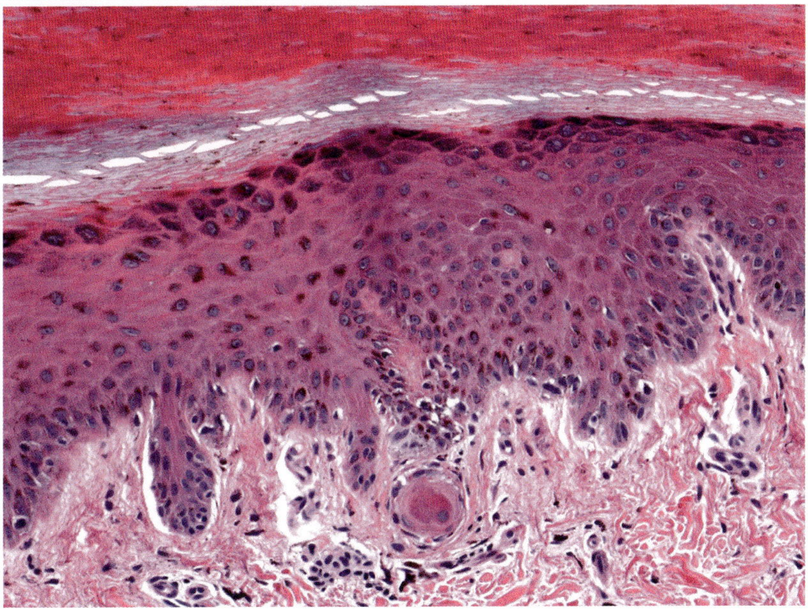

Figs. 7.9 and 7.10 Even at high-power view, diagnostic features were quite subtle. Scattered atypical, dendritic melanocytes, always basally located, could be identified, but pleomorphism was minimal. Lymphocytic infiltrate, a feature often described in acral melanoma, was absent

Case 3

Patient: Female, 50 years old

Anatomic site of the lesion: Acral, left foot

History of the lesion: Unknown

Management: The clinical and dermoscopic images were highly suggestive for melanoma.

Histopathologic diagnosis: Junctional nevus in acral skin. As clinical and dermoscopic features were a cause of concern, some doubts still remained on the risk of underdiagnosing a very early melanoma in situ.

Key message: Like other fields in the pathology of pigmented neoplasias, melanocytic lesions on the acral sites carry the risk to be underdiagnosed in order not to be overdiagnosed.

In the present case, clinical features were somehow discordant from histological findings, the latter being extremely subtle and not sufficient to support a diagnosis of in situ melanoma. However, a pragmatical approach in case of uncertainty, that is, the excision with wide margins, was needed for the patient's safety.

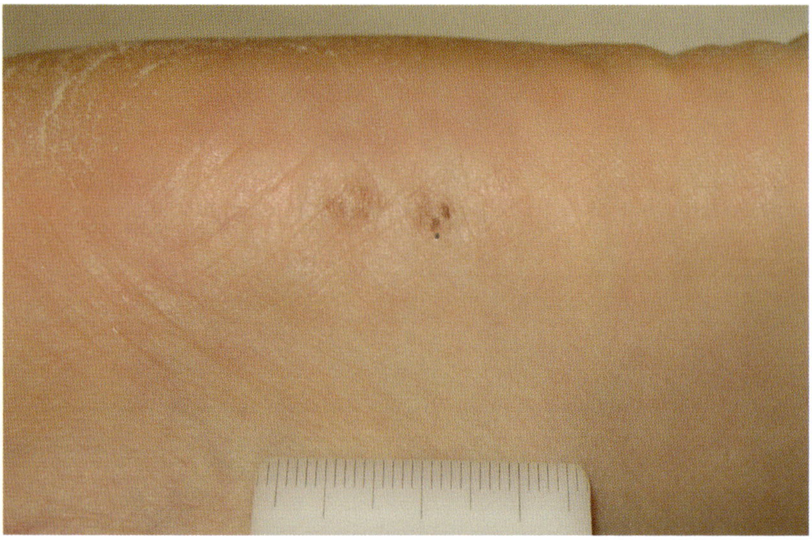

Fig. 7.11 The clinical examination revealed a solitary asymmetric large pigmented macule

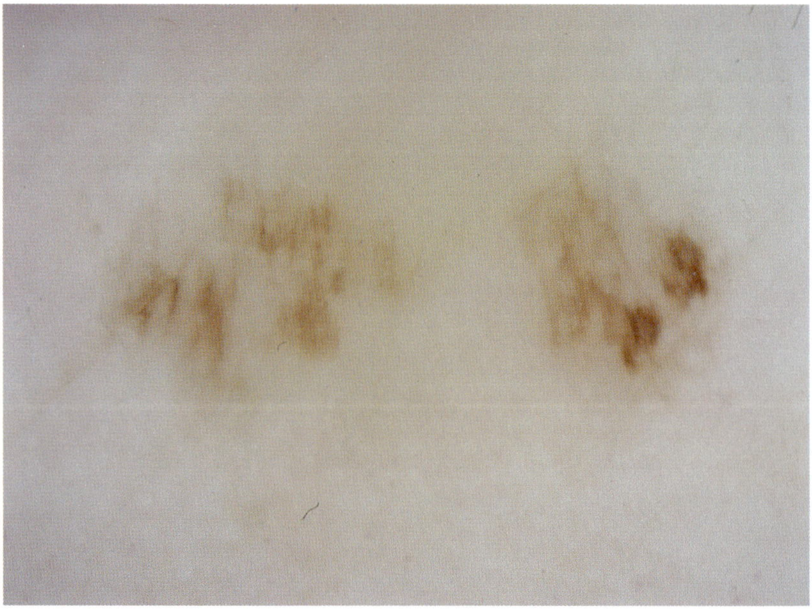

Fig. 7.12 Dermoscopy revealed a parallel pattern with color ranging from light to dark brown and asymmetry for shape and color distribution

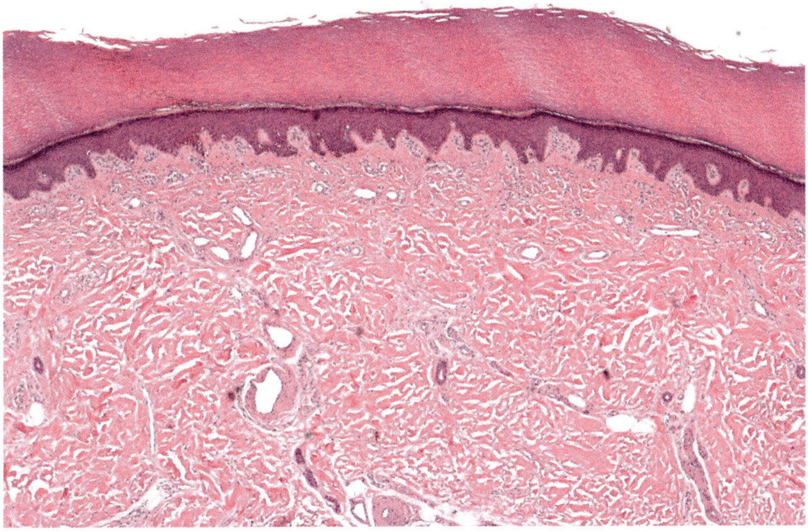

Fig. 7.13 At scanning magnification, the lesion was limited to few basally located melanocytes. The epidermis was not hyperplastic

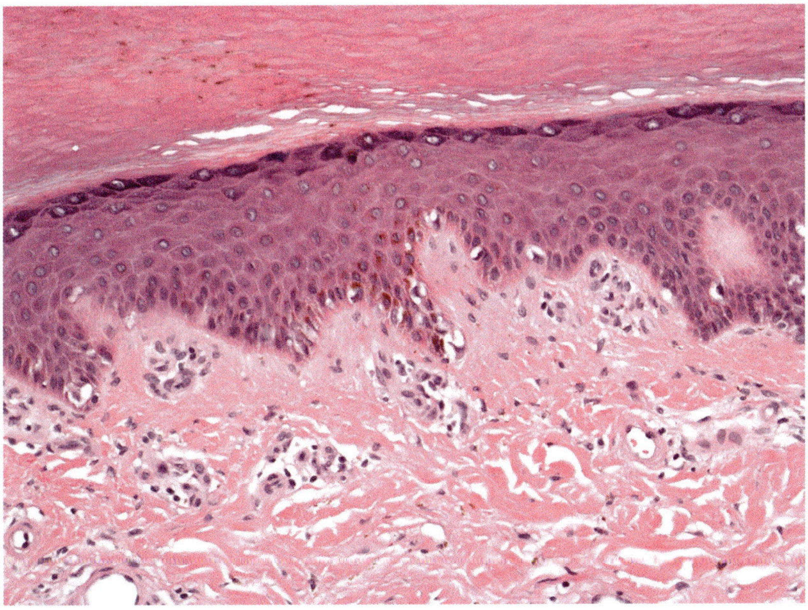

Fig. 7.14 At higher magnification, some melanocytes were present in single units. Atypia was not a cytological feature nor pagetoid spread

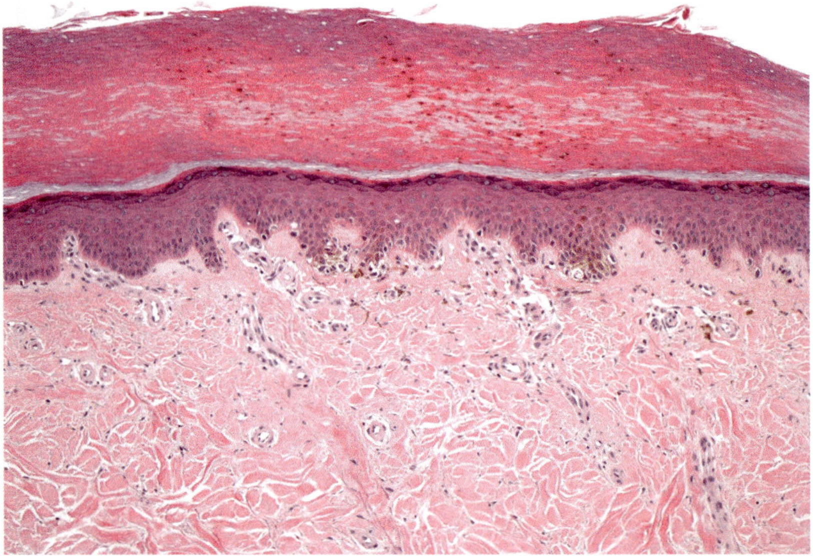

Fig. 7.15 Occasionally, proliferating melanocytes were arranged in small nests showing slight atypia

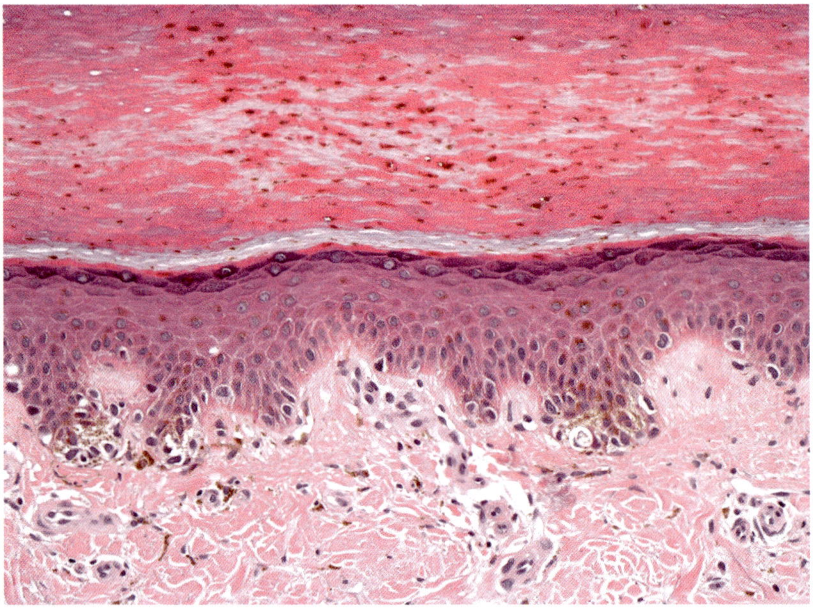

Fig. 7.16 The relatively paucity of melanocytes makes the clinicopathological correlation necessary to render the diagnosis

Suggested Reading

1. Fernandez-Flores A, Cassarino DS. Histopathological diagnosis of acral lentiginous melanoma in early stages. Ann Diagn Pathol. 2017;26:64–9.
2. Durbec F, Martin L, Derancourt C, Grange F. Melanoma of the hand and foot: epidemiological, prognostic and genetic features. A systematic review. Br J Dermatol. 2012;166(4):727–39.
3. Bravo Puccio F, Chian C. Acral junctional nevus versus acral lentiginous melanoma in situ: a differential diagnosis that should be based on clinicopathologic correlation. Arch Pathol Lab Med. 2011;135(7):847–52.
4. Lallas A, Kyrgidis A, Koga H, Moscarella E, Tschandl P, Apalla Z, Di Stefani A, Ioannides D, Kittler H, Kobayashi K, Lazaridou E, Longo C, Phan A, Saida T, Tanaka M, Thomas L, Zalaudek I, Argenziano G. The BRAAFF checklist: a new dermoscopic algorithm for diagnosing acral melanoma. Br J Dermatol. 2015;173(4):1041–9.
5. Ahn CS, Guerra A, Sangüeza OP. Melanocytic nevi of special sites. Am J Dermatopathol. 2016;38(12):867–81.
6. Hosler GA, Moresi JM, Barrett TL. Nevi with site-related atypia: a review of melanocytic nevi with atypical histologic features based on anatomic site. J Cutan Pathol. 2008;35(10):889–98.
7. Kim JY, Choi M, Jo SJ, Min HS, Cho KH. Acral lentiginous melanoma: indolent subtype with long radial growth phase. Am J Dermatopathol. 2014;36(2):142–7.

8 Melanoma Incognito

The term "melanoma incognito" refers to a clinically false-negative melanoma mimicking a benign lesion, such as harmless hemangioma, seborrheic keratosis, common intradermal nevus, or even an inflammatory lesion. It has also been called "featureless melanoma" to emphasize that clinical examination does not allow to reach a correct diagnosis as this rare type of melanoma does not exhibit a sufficient array of clinical features to justify biopsy. Interestingly, when biopsied, such cases do not cause difficulties in dermatopathological diagnosis, being usually histologically clear-cut melanomas, superficial spreading, nevoid, or even spitzoid.

As melanoma incognito has a significant potential to pose a great threat to the patient affected, clinicians must consider that a clinically banal or borderline lesion might in fact be a melanoma.

Dermoscopy is routinely used as a second-level diagnostic procedure to screen all pigmented lesions in a given patient. In fact, a total body skin check is routinely made with dermoscopy for clinically looking lesions to avoid missing an occult melanoma and for clinically suspect lesions, thereby decreasing the number of unnecessary biopsies of benign lesions.

To avoid missing a melanoma incognito, some golden rules can be followed in clinical practice.

First of all, dermoscopy should not only be used for suspicious skin lesions. In fact, all pigmented lesions in a given patient should be carefully screened by dermoscopy regardless of their clinical appearance. Second, we need to biopsy all lesions missing good clinico-dermoscopic correlation. Third, another golden rule is to biopsy lesions with nonspecific global pattern and lesions that deviate from the patients' other typical-looking nevi. Forth, clinicians should analyze carefully pigmented lesion changes tracked through digital imaging. Fifth, if a clinician sees pink color clinically or with dermoscopy, he should raise a red flag. In pink lesions, the analysis of vascular patterns is very important and can aid in the dermoscopic differential diagnosis especially when the atypical vascular pattern consists of dotted and linear-irregular vessels.

Sixth, the presence of regression structures is another dermoscopic clue that should be kept in mind to avoid missing melanoma, especially when evaluating patients with multiple acquired melanocytic nevi.

In conclusion, if one looks at this type of lesion with dermoscopy, there might be clues that raise one's index of suspicion enough to warrant a histopathologic diagnosis.

C. Longo et al., *Atlas of Diagnostically Challenging Melanocytic Neoplasms*,
https://doi.org/10.1007/978-3-319-48653-6_8

Case 1

Patient: Male, 79 years old

Anatomic site of the lesion: posterior left leg

History of the lesion: unknown.

Management: any nodular tumor should be promptly excised, especially in the suspicion of squamous cell carcinoma.

Histopathologic Diagnosis: nodular melanoma

Key message: Diagnosing a nodular melanoma is challenging because it can mimic any benign or other malignant entities. This is even more true in case of amelanotic melanoma when few dermoscopic clues are present. In this case, the presence of white color associated to atypical vascular pattern was misleading towards the diagnosis of epithelial tumor. However, it is well known that the presence of atypical vascular pattern is a common finding in any aggressive tumors such as melanoma or squamous cell carcinoma. Furthermore, looking a posteriori, it is possible to note a tiny brown color at the edge of the nodule that could be suggestive for a melanocytic proliferation. As a general rule, a prompted excision should be done in case of nodules with atypical vascular pattern. In adjunct to dermoscopy, confocal microscopy could be of help to recognize a melanoma.

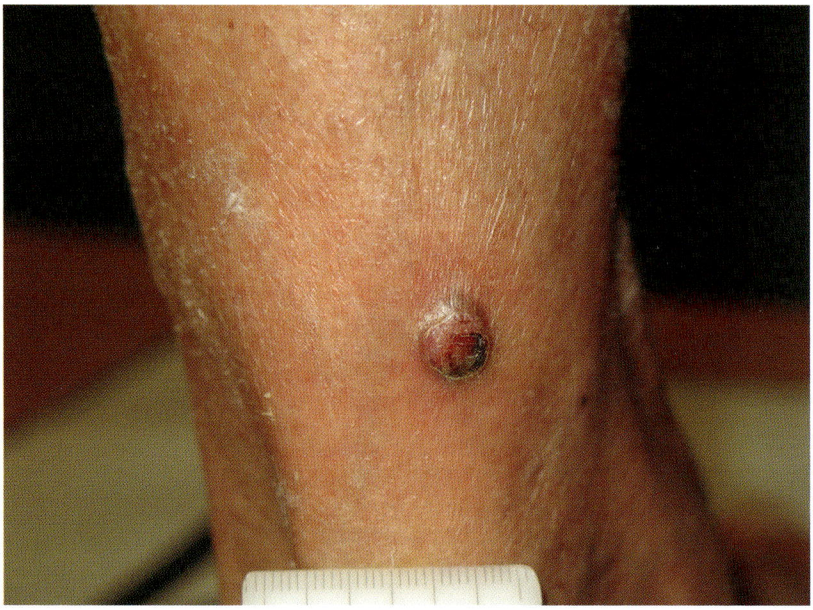

Fig. 8.1 Solitary ulcerated nodular lesion

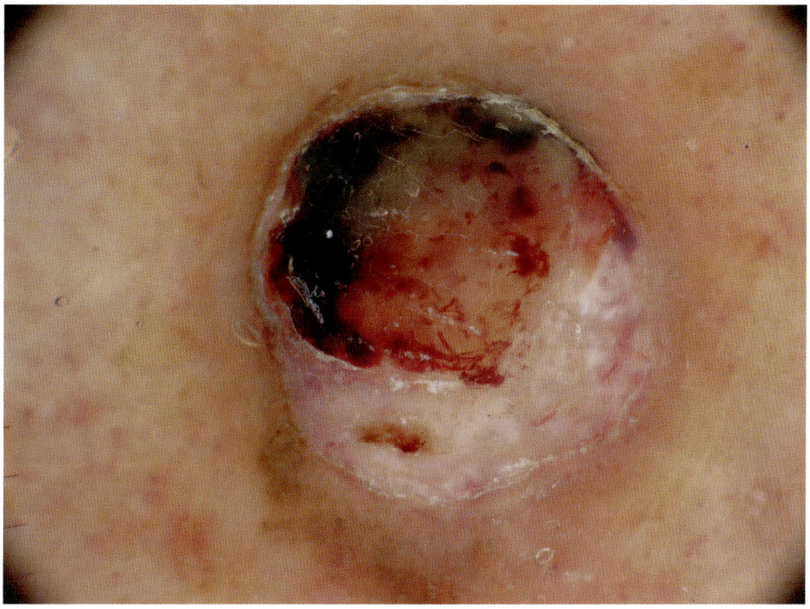

Fig. 8.2 Dermoscopy shows an ulcerated nodule with white areas and polymorphous vessels. Overall, the lesion could be provisionally classified as squamous cell carcinoma

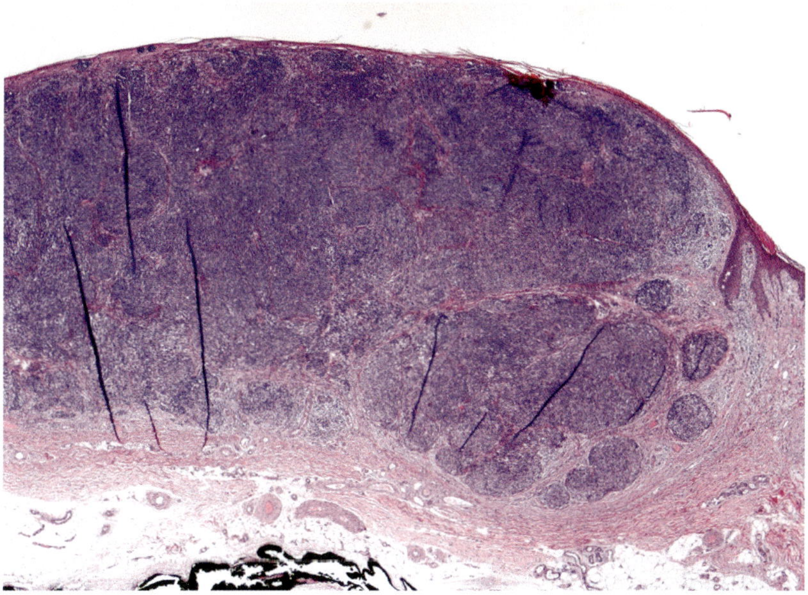

Fig. 8.3 The tumor was nodular and very cellular, composed of spindle melanocytes, deeply invading the dermis

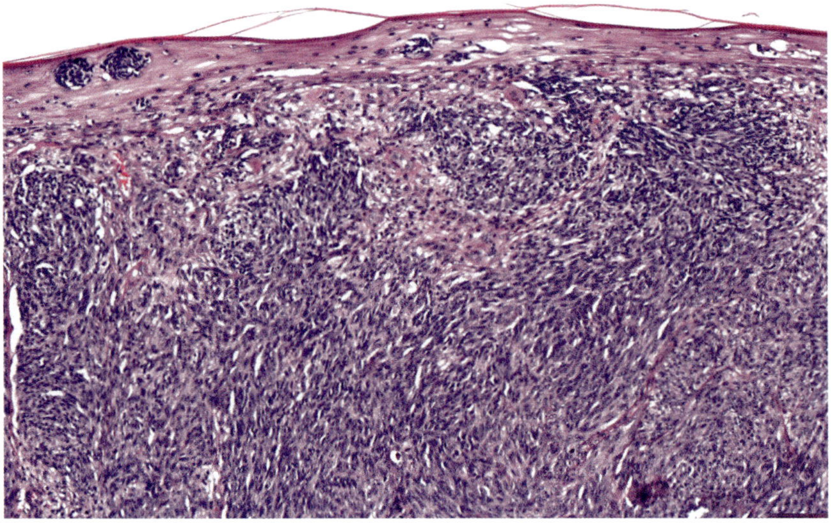

Fig. 8.4 Melanocytes are spindle, arranged in fascicles and with no sign of maturation and pigmentation. An immunohistochemical stain with MART-1 confirmed the diagnosis

Case 2

Patient: Female, 28 years old

Anatomic site of the lesion: Left upper limb

History of the lesion: Unknown; the patient had an invasive spitzoid melanoma on her foot 3 years ago.

Management: The lesion was in differential diagnosis with melanoma and thus was excised.

Histopathologic diagnosis: Nevoid melanoma

Key message: Nevoid melanoma, a rare variant of melanoma, can be difficult to detect as its clinical and histopathological morphology can simulate a nevus. However, the presence of atypical vessels and asymmetric distribution of pigment should be considered as suspicious for melanoma diagnosis.

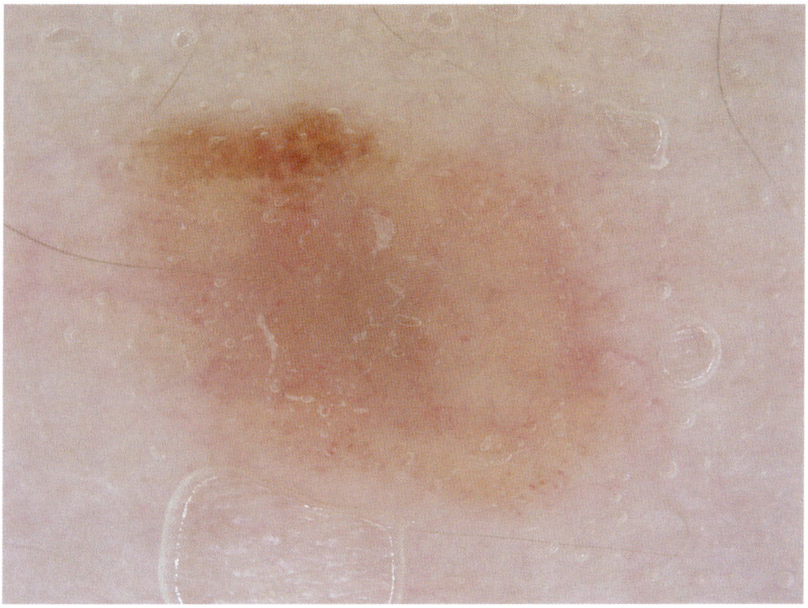

Fig. 8.5 Dermoscopically, the lesion was asymmetric: the majority showed a structureless pinkish area with barely visible dotted vessels, whereas on the peripheral left side, an inverse light-brown network was visible

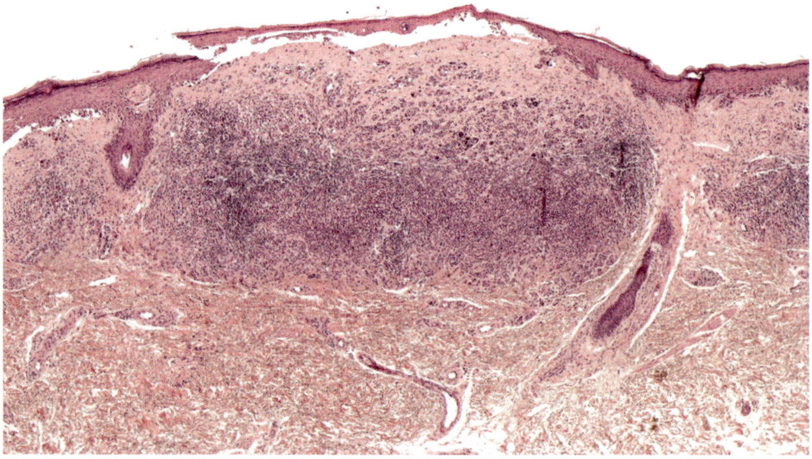

Fig. 8.6 At low-power view, the lesion was an achromic, relatively ill-defined papule

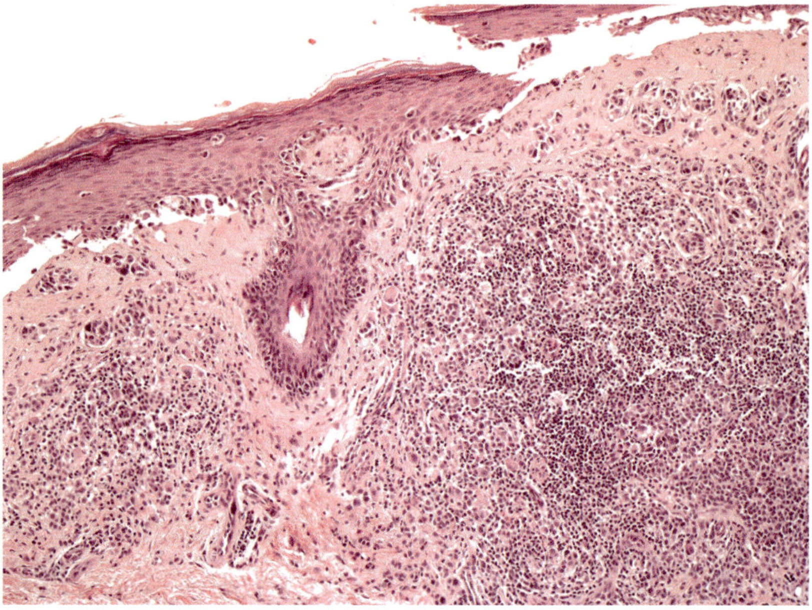

Fig. 8.7 Cytologically, it was composed of monomorphic, hyperchromatic small melanocytes, admixed with a brisk lymphocytic inflammation. Few mitotic figures can be found within the lesion

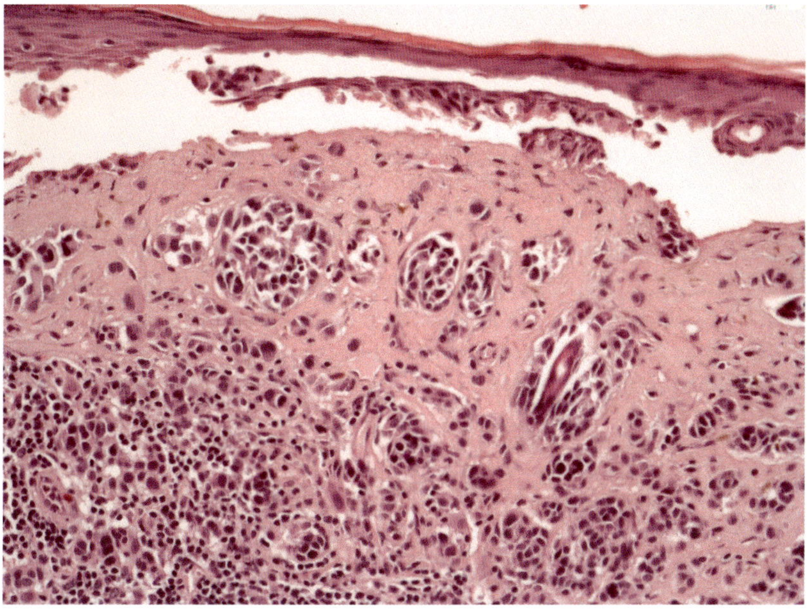

Fig. 8.8 The epidermal infiltration, evident at the periphery of the papule, resulted in an artefactual dermoepidermal detachment

Case 3

Patient: Female, 22 years old

Anatomic site of the lesion: Left cheek

History of the lesion: The lesion appeared few weeks ago and rapidly enlarged.

Management: The differential diagnosis included a spitzoid lesion and a vascular tumor. The presence of atypical vessels including serpentine, linear-irregular, and looped ones was not compatible with the diagnosis of angioma, which prompted us to excise the lesion to rule out spitzoid melanoma.

Histopathologic diagnosis: Melanoma with spitzoid, pleomorphic cells

Key message: Red nodules are difficult to diagnose. As a general rule, it is mandatory to look first at possible benign dermoscopic patterns. In this case, no dermoscopic features for angioma or pyogenic granuloma were noted. Furthermore, an atypical vascular pattern was noted along with the history of a new fast-growing lesion. The presence of one single dermoscopic criterion and the absence of benign features are enough to decide that the lesion could be a melanoma.

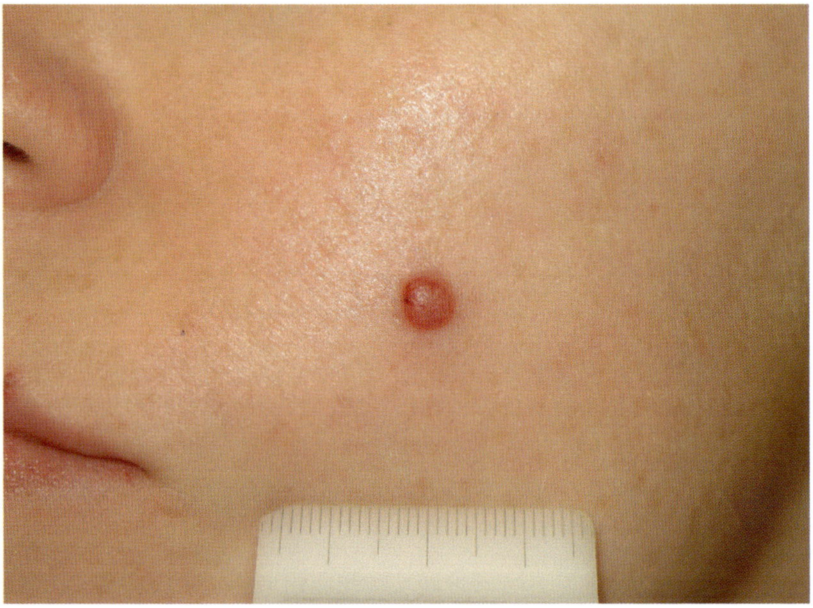

Fig. 8.9 Clinical close-up examination revealed a symmetric roundish small red nodule without evidence of pigmentation

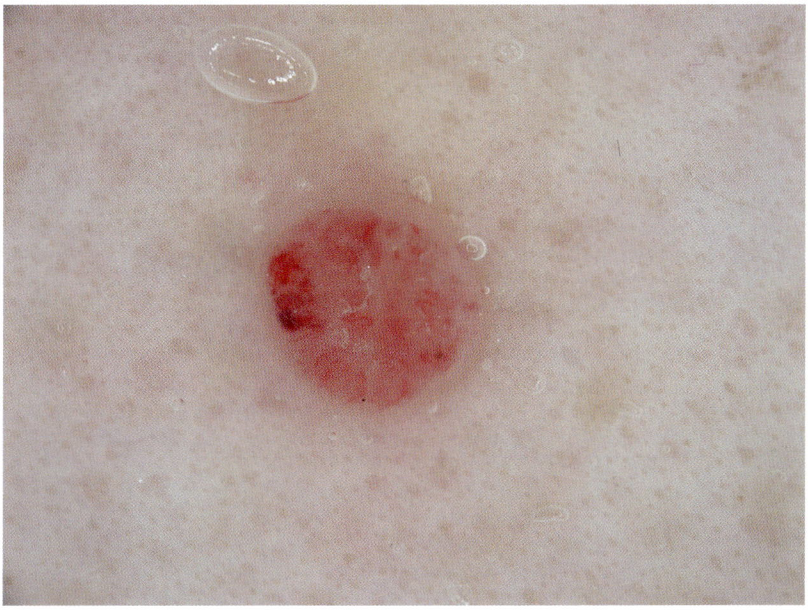

Fig. 8.10 Dermoscopy revealed multiple atypical vessels over a reddish background

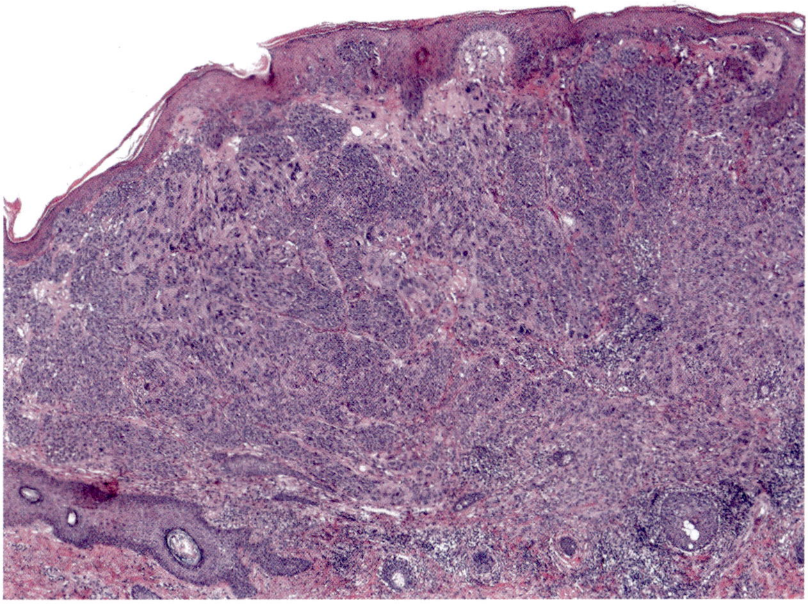

Fig. 8.11 At low power, the lesion was a well-defined nodule made up of pleomorphic cells

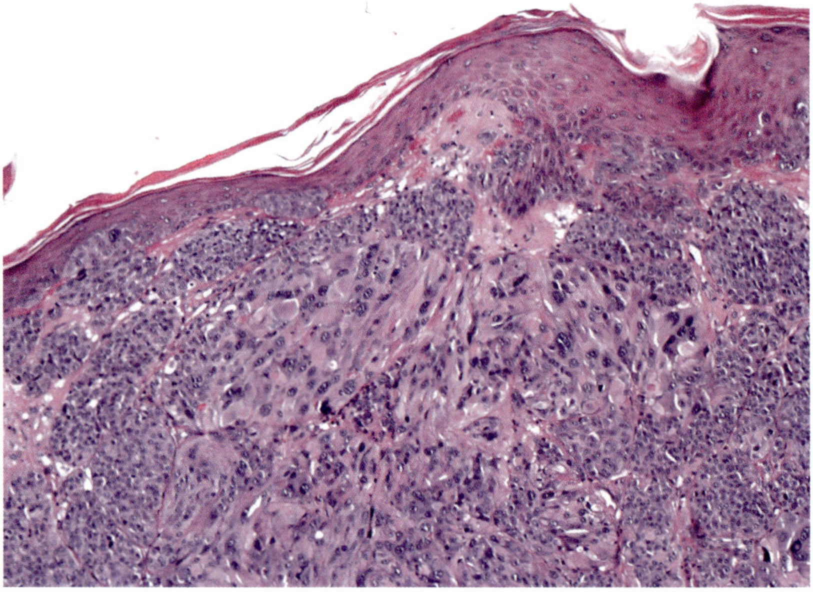

Fig. 8.12 Large epithelioid melanocytes were intermingled with small elements. The epidermis was only focally involved

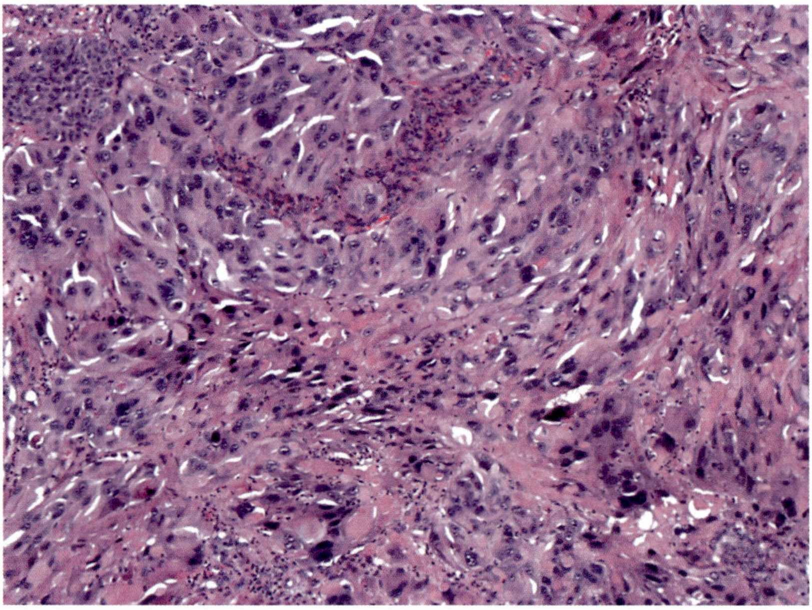

Fig. 8.13 Cytological atypia was striking and mitotic figures were brisk. An immunohistochemical stain with MART-1 confirmed the diagnosis

Case 4

Patient: Male, 29 years old

Anatomic site of the lesion: Right scapula

History of the lesion: The patient had several moles on the trunk and the extremities. Among them, there was one newly appearing translucid papule.

Management: The lesion was excised to rule out melanoma.

Histopathologic diagnosis: Invasive melanoma, with features consistent with nevoid melanoma

Key message: Despite the overall symmetry of the lesion, the presence of inverse network and bluish color (i.e., veil) favored the diagnosis of melanoma although a spitzoid lesion could not be excluded with certainty. In this case, histopathology revealed a nevoid component that did not find a specific dermoscopic correlate.

However, the risk to miss a melanoma is quite low since the golden rule is to excise any given lesion with those dermoscopic aspects seen in the present case.

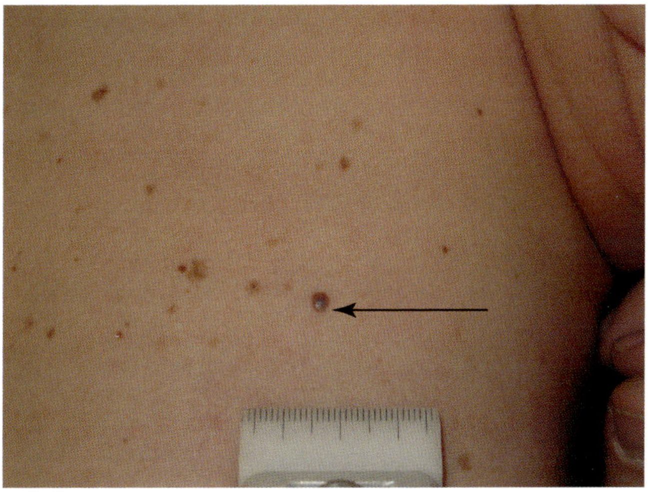

Fig. 8.14 Close-up clinical examination revealed a roundish brown papule with a diameter of 0.4 cm, with regular borders

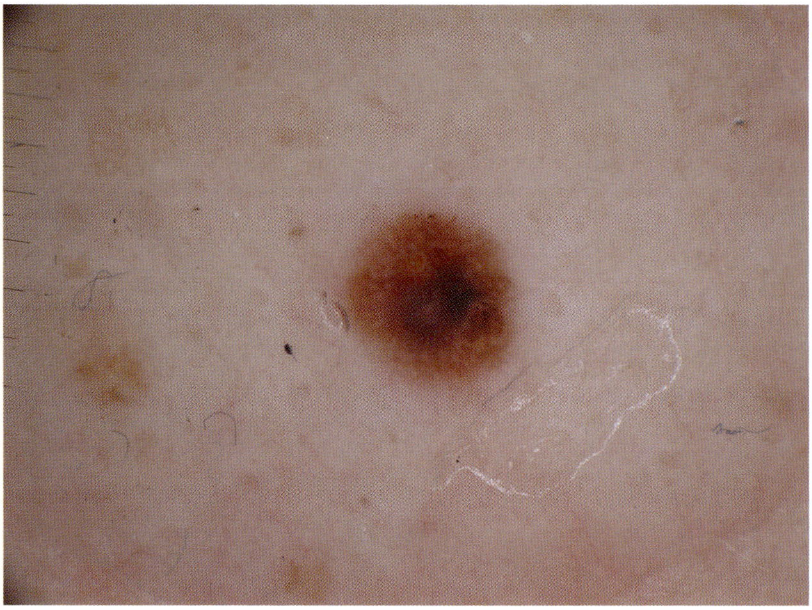

Fig. 8.15 Dermoscopically, the lesion was characterized by the presence of an inverse network with irregularly distributed globules and dotted vessels intersected by white lines. A bluish area was visible in the center of the lesion

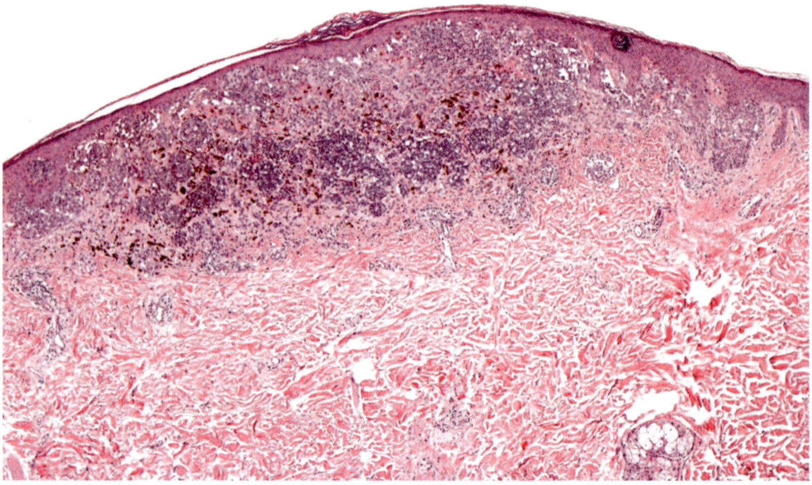

Fig. 8.16 The lesion was an asymmetrical papule with a "busy," highly cellular dermis

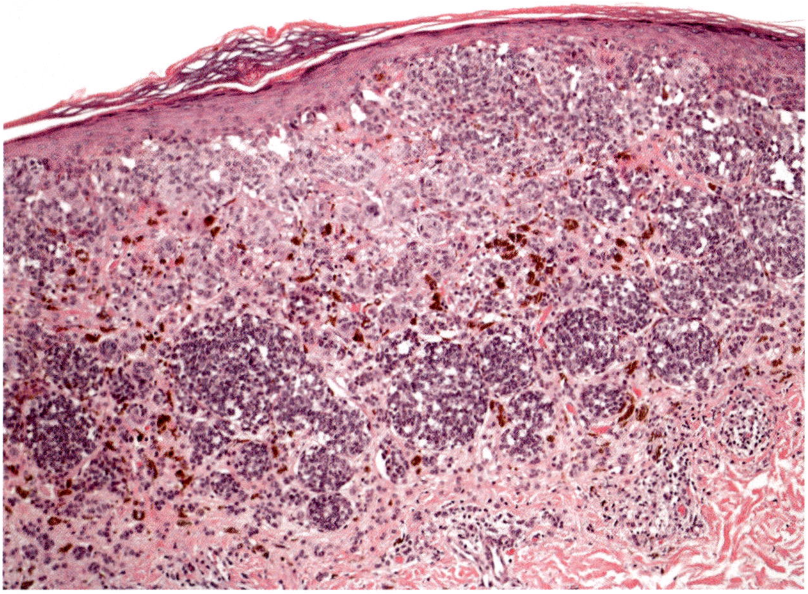

Fig. 8.17 At higher magnification, melanocytes were atypical and densely packed. The epidermis was unremarkable

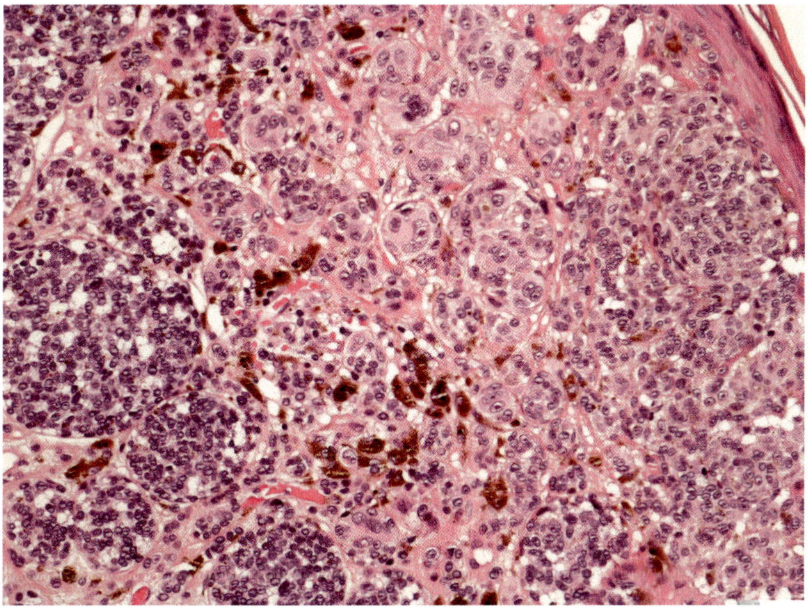

Fig. 8.18 In this field, intratumoral cellular heterogeneity was evident; small cells with scant cytoplasm and blue nuclei were juxtaposed to epithelioid elements with abundant eosinophilic cytoplasm

Case 5

Patient: Female, 42 years old

Anatomic site of the lesion: Left lower leg

History of the lesion: Appeared few months ago

Management: Because of the bluish color and the lack of clear-cut benign features, the lesion was excised to rule out melanoma

Histopathologic diagnosis: The differential diagnosis was between a pagetoid Spitz nevus and in situ melanoma. Some morphologic features, mainly the pagetoid spread at the border of the lesion, favored the latter hypothesis.

Key message: The pagetoid Spitz nevus is a rare variant of Spitz nevus, characterized by intraepidermal proliferation of epithelioid melanocytes. They are mostly arranged in single units and tend to the upward migration through the epidermis (the so-called pagetoid pattern), thus mimicking melanoma in situ. In addition, female thighs have been claimed as a special site since pagetoid features are frequently observed even in otherwise benign lesions.

The present case, even though recapitulating these differential diagnoses, carried cytomorphologic features convincing for melanoma in situ. The history of being a new lesion is not relevant since both melanoma and Spitz nevi share the same attitude of growth. However, dermoscopically, this lesion was not typified by the starburst or globular pattern, and thus the diagnosis of melanoma was more likely.

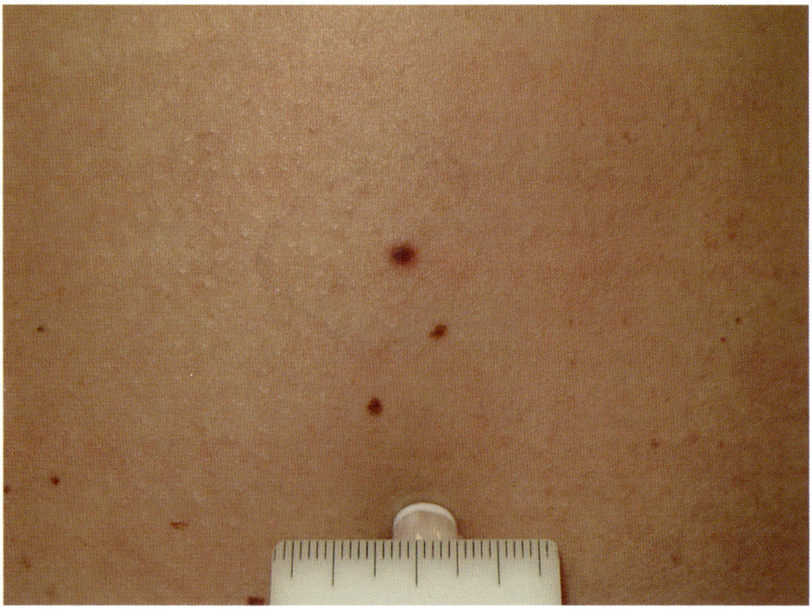

Fig. 8.19 An elevated roundish papule of light-brown color with reddish hue with a diameter of 4 mm. Other pigmented lesions were noted in proximity of this newly appearing lesion

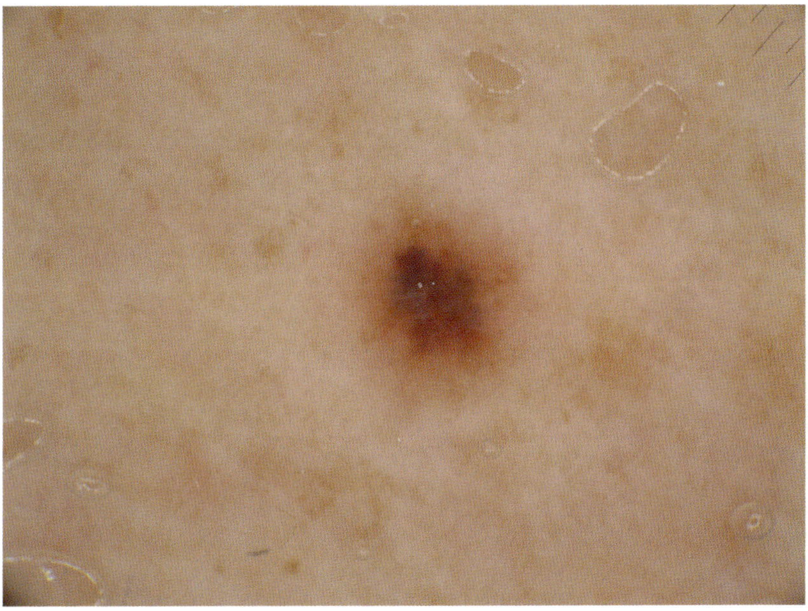

Fig. 8.20 Dermoscopy revealed a structureless bluish central area with fading off light-brown peripheral network

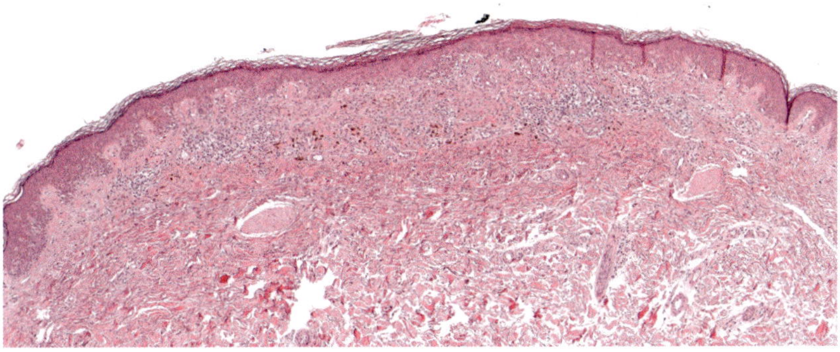

Fig. 8.21 Histologically the lesion was very thin but ill defined, with no definite lateral borders

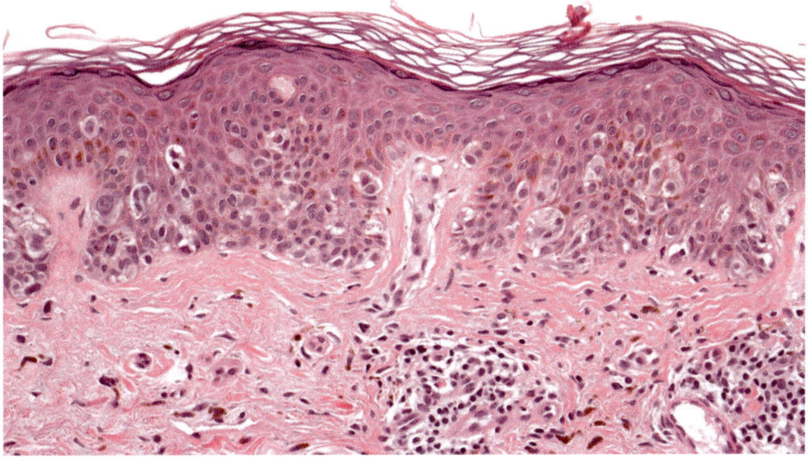

Fig. 8.22 At high-power view, melanocytes were epithelioid and showed the tendency to migrate upward in the epidermis

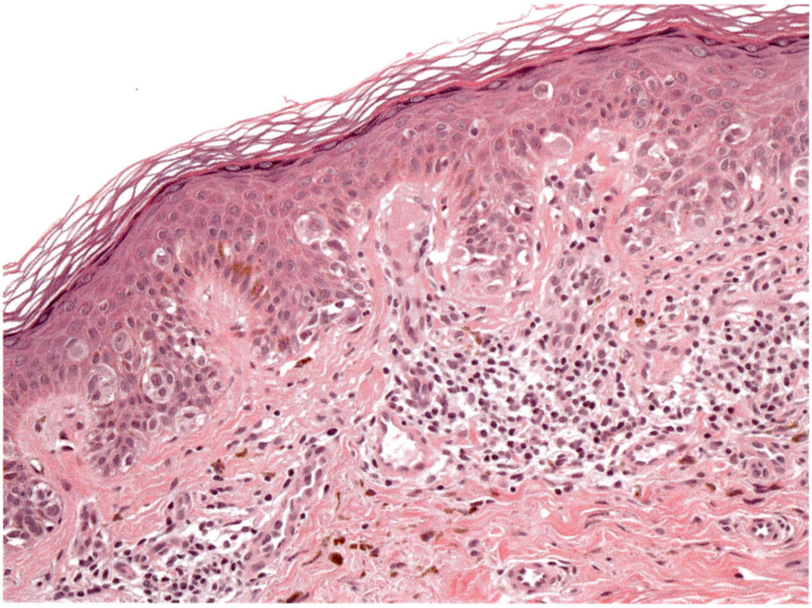

Fig. 8.23 Also at the periphery of the lesions, melanocytes showed pagetoid spread through the epidermis. They were also slightly pleomorphic

Suggested Reading

1. Shitara D, Nascimento M, Ishioka P, Carrera C, Alós L, Malvehy J, Puig S. Dermoscopy of Naevus-associated melanomas. Acta Derm Venereol. 2015;95(6):671–5.
2. Haenssle HA, Mograby N, Ngassa A, Buhl T, Emmert S, Schön MP, Rosenberger A, Bertsch HP. Association of patient risk factors and frequency of nevus-associated cutaneous melanomas. JAMA Dermatol. 2016;152(3):291–8.
3. Longo C, Piana S, Marghoob A, Cavicchini S, Rubegni P, Cota C, Ferrara G, Cesinaro AM, Baade A, Bencini PL, Mantoux F, Mijuskovic ZP, Pizzichetta MA, Stanganelli I, Carrera C, Giovene GL, Ranasinghe S, Zalaudek I, Lallas A, Moscarella E, Specchio F, Pepe P, Pellacani G, Argenziano G. Morphological features of naevoid melanoma: results of a multicentre study of the International Dermoscopy Society. Br J Dermatol. 2015;172(4):961–7.
4. Diwan AH, Lazar AJ. Nevoid melanoma. Clin Lab Med. 2011;31(2):243–53.
5. Idriss MH, Rizwan L, Sferuzza A, Wasserman E, Kazlouskaya V, Elston DM. Nevoid melanoma: a study of 43 cases with emphasis on growth pattern. J Am Acad Dermatol. 2015;73(5):836–42.
6. Argenziano G, Zalaudek I, Ferrara G, Johr R, Langford D, Puig S, Soyer HP, Malvehy J. Dermoscopy features of **melanoma incognito**: indications for biopsy. J Am Acad Dermatol. 2007;56(3):508–13.
7. Fernandez AP, Billings SD, Bergfeld WF, Ko JS, Piliang MP. Pagetoid Spitz nevi: clinicopathologic characterization of a series of 12 cases. J Cutan Pathol. 2016;43(11):932–9.
8. Gerami P, Barnhill RL, Beilfuss BA, LeBoit P, Schneider P, Guitart J. Superficial melanocytic neoplasms with pagetoid melanocytosis: a study of interobserver concordance and correlation with FISH. Am J Surg Pathol. 2010;34(6):816–21.
9. Requena C, Requena L, Kutzner H, Sanchez YE. Spitz nevus: a clinicopathological study of 349 cases. Am J Dermatopathol. 2009;31(2):107–16.
10. Buonaccorsi JN, Lynott J, Plaza JA. Atypical melanocytic lesions of the thigh with spitzoid and dysplastic features: a clinicopathologic study of 29 cases. Ann Diagn Pathol. 2013;17(3):265–9.
11. Donati P, Muscardin L, Cota C, Panetta C, Paolino G. The melanocytic epithelioid cell nevus of the thigh of woman: a simulator of melanoma. Am J Dermatopathol. 2012;34(8):853–5.
12. Lallas A, Moscarella E, Longo C, et al. Likelihood of finding melanoma when removing a spitzoid-looking lesion in patients aged 12 years or older. J Am Acad Dermatol. 2015;72:47–53.
13. Menzies SW, Moloney FJ, Byth K, Avramidis M, Argenziano G, Zalaudek I, Braun RP, Malvehy J, Puig S, Rabinovitz HS, Oliviero M, Cabo H, Bono R, Pizzichetta MA, Claeson M, Gaffney DC, Soyer HP, Stanganelli I, Scolyer RA, Guitera P, Kelly J, McCurdy O, Llambrich A, Marghoob AA, Zaballos P, Kirchesch

HM, Piccolo D, Bowling J, Thomas L, Terstappen K, Tanaka M, Pellacani G, Pagnanelli G, Ghigliotti G, Ortega BC, Crafter G, Ortiz AM, Tromme I, Karaarslan IK, Ozdemir F, Tam A, Landi C, Norton P, Kaçar N, Rudnicka L, Slowinska M, Simionescu O, Di Stefani A, Coates E, Kreusch J. Dermoscopic evaluation of nodular melanoma. JAMA Dermatol. 2013;149(6):699–709.

14. Lallas A, Zalaudek I, Apalla Z, Longo C, Moscarella E, Piana S, Reggiani C, Argenziano G. Management rules to detect melanoma. Dermatology. 2013;226(1):52–60.

15. Zalaudek I, Kreusch J, Giacomel J, Ferrara G, Catricalà C, Argenziano G. How to diagnose nonpigmented skin tumors: a review of vascular structures seen with dermoscopy: part II. Nonmelanocytic skin tumors. J Am Acad Dermatol. 2010;63(3):377–86.

16. Longo C, Farnetani F, Ciardo S, Cesinaro AM, Moscarella E, Ponti G, Zalaudek I, Argenziano G, Pellacani G. Is confocal microscopy a valuable tool in diagnosing nodular lesions? A study of 140 cases. Br J Dermatol. 2013;169(1):58–67.

Printed by Printforce, the Netherlands